Department of Education and Science
Welsh Office
Department of Education for Northern Ireland

Assessment of Performance Unit

Science at Age 15

A Review of APU Survey Findings 1980-84

Fred Archenhold (Editor)
John Bell
James Donnelly
Sandra Johnson
Geoff Welford

London
Her Majesty's Stationery Office

First published 1988

ISBN 0 11 270615 0

Contents and authorship

Editor: Fred Archenhold

Tables and figures

Tables

Figures

Preface

This report presents a review of the findings of five national surveys on the performance of 15 year old children in science*. The surveys were conducted for the Assessment of Performance Unit (APU) by research teams at King's College, University of London, and at the University of Leeds. The report gives an overview of the initial series of five annual surveys of this age, carried out between 1980 and 1984. Surveys will in future take place at five-yearly intervals. Companion reports are also available for ages 11 and 13, together with a technical report which considers issues of interpretation raised by the monitoring programme ('National assessment: the APU science approach', Johnson, 1988).

This report describes in detail the approach to the surveys and the findings. It is intended primarily for administrators, education advisers and researchers, but it will also be of interest to those teachers who found the previous reports of value. Further copies of this and the other review reports may be purchased from HMSO or through booksellers. A full list of APU publications and details of their availability can be obtained from: APU, Room 4/26, Department of Education and Science, Elizabeth House, York Road, London SE1 7PH (telephone: 01-934 9323).

Details of short reports for science teachers drawing on the findings of the APU surveys are available from the above address.

Reports now available in the series:

Number 1	Science at Age 11
Number 2	Science Assessment Framework Ages 13 and 15
Number 3	Science at Age 13
Number 4	Science Assessment Framework Age 11
Number 5	Science at Age 15
Number 6	Practical Testing at Ages 11, 13 and 15
Number 7	Electricity at Age 15
Number 8	Planning Scientific Investigations at Age 11
Number 9	Assessing Investigations in Science at Ages 13 and 15
Number 10	Metals at Age 15
Number 11	The Language of Science

---

* Since the testing reviewed in this report was carried out there have been significant educational developments which will have considerable effects on the teaching of science in schools: i) the GCSE has been introduced (starting Autumn 1986) to replace GCE O-levels, CSE and Joint O-level/CSE examinations; ii) the Education Reform Bill (currently before Parliament) has proposed the introduction of a national curriculum for school children which will, *inter alia*, make the study of science compulsory until age 16, with regular assessments of performance at ages 7, 11, 14 and 16.

The research teams

The teams are based at two centres, at the University of Leeds and at King's College London (formerly Chelsea College). The work of monitoring at three ages (11, 13 and 15) in science is divided between these centres so that the responsibility for monitoring at the lower age groups rests with King's and the responsibility for monitoring at age 15 and for the data processing and analysis rests with Leeds.

The science teams (*January 1987*)

Leeds:

Director	Fred Archenhold
Technical Director	Roger Hartley
Research and Development (age 15)	Geoff Welford (Project Coordinator) James Donnelly
Data Analysis	Sandra Johnson (Deputy Director) John Bell Mark Tranmer
Secretary	Jan Akkermans

King's:

Director	Paul Black
Research and Development (ages 11 and 13)	Patricia Murphy (Deputy Director) Peter Swatton Robert Taylor Anne Qualter
Secretary	Julie Jones

Past team members

Dennis Child (Director 1982–83); Angela Davey (1982–84); Brenda Denvir (1981); Rosalind Driver (1977–82, Deputy Director 1979–82); Reed Gamble (1982–85); Richard Gott (1980–84, Deputy Director 1982–84); Wynne Harlen (1977–1984, Deputy Director (1978–84); Jenny Head (1981–84); Nasrin Khaligh (1984–85); David Layton (Director 1977–82); Brian Maher (1979–83); Cynthia Millband (1980–81); Tony Orgee (1982–85); David Palacio (1980–85); Terry Russell (1982–86); Beta Schofield (1977–85, Deputy Director 1985); Karen Spencer (1985–86); Ardrie VanderWaal (1979–80); Iain Watson (1984–85); Christopher Worsley (1978–82); Fiona Wylie (1980–82).

Acknowledgements

The monitoring team would like to thank the many people who contributed to the various phases of work involved in conducting the five annual surveys and producing this review report.

The monitoring framework and test questions used were the result of a number of years of development work cooperatively undertaken by the teams at King's College London (formerly Chelsea College) and at the University of Leeds in collaboration with the Science Steering Group.

Mrs B. A. Bloomfield and her staff at the Monitoring Services Unit were responsible for obtaining the co-operation of the schools drawn in the sample and making all the administrative arrangements with Heads and teachers for both written and practical testing.

Science advisers in the local education authorities helped by nominating teachers as practical testers and arranging for their release from normal teaching duties. Dr N. B. Evans, HMI, arranged for the translation of test materials into Welsh. In Northern Ireland, Mr I. W. Milligan, HMI, and in Wales, Dr N. B. Evans, HMI, made arrangements for the recruitment of testers.

Philip Harris Biological Ltd supplied the apparatus for the practical testing. The care and ingenuity of members of their staff have been greatly appreciated.

The task of marking the written packages and administering the practical tests during 1980–84 was undertaken most competently by science teachers from England, Wales and Northern Ireland.

The conduct of the surveys themselves depended on the goodwill and cooperation of pupils and staff in the survey schools. We are particularly grateful to the teachers, who undertook the various time-consuming tasks involved in administering the tests, and without whose help these surveys would not have been possible.

The Leeds University members of the present science team, who have had the responsibility of writing this review report, wish to acknowledge the invaluable contributions made to all aspects of APU research, including the surveys, by present and past colleagues based at King's College London, and at age 15, in particular, by past team members at the University of Leeds: Dennis Child, Angela Davey, Ros Driver, Reed Gamble, Richard Gott, Jenny Head, Nasrin Khaligh, David Layton, Brian Maher, Iain Watson, Christopher Worsley and Fiona Wylie.

The project secretarial staff, Nadine Hannam, Helen Johnson, Elizabeth Lodge, Jan Akkermans and, in particular, Glynis Wilkinson, deserve special thanks for the efficient manner in which they have assisted in the production of all the reports.

1

Overview of APU science surveys 1980–84

1.1 The purpose of this review report

This review report draws together, in summary form, the findings of the five annual surveys of the science performance of 15 year olds conducted in England, Wales and Northern Ireland in the period 1980–84. Two companion review reports summarise the findings for pupils aged 11 and 13 respectively. A technical report (Johnson, 1988) considers the measurement and interpretational issues raised by this innovative and large scale national monitoring programme.

The APU science monitoring project was designed to provide evidence of levels of performance of the population at three ages. Full reports have been published of the performance of 15 year old pupils in the surveys conducted in 1980 (Driver *et al*, 1982), in 1981 (Driver *et al*, 1984), in 1982 (Gott *et al*, 1985) and in 1983 (Welford *et al*, 1986). Each of these reports gives a detailed account of the survey conducted in a particular year with examples of the types of questions used. Page references are made to these reports, as appropriate, to enable points of detail or specific test questions to be easily located. The results of the 1984 survey are included in this review report.

The next national surveys of 13 and 15 year old pupils in science are planned to take place in 1989 and of 11 year old pupils in 1990. In the meantime, existing data from the annual surveys in the period 1980–84 are being subjected to further analysis and a number of in-depth research studies are in progress to address interpretational issues arising from the APU science results. It is intended that the dissemination of these studies should supplement this set of review reports and the 'Science reports for teachers' published by the Assessment of Performance Unit and available from the Association for Science Education. A list of 'Science reports for teachers' published so far is included in Appendix 10.

1.2 The development of the science monitoring programme

The Assessment of Performance Unit was set up by the Department of Education and Science in 1975 'to promote the development of methods of assessing and monitoring the achievement of children at school, and to seek to identify the incidence of under-achievement'. Associated with these terms of reference were four tasks:

- to identify and appraise existing instruments and methods of assessment which may be relevant for these purposes;
- to sponsor the creation of new instruments and techniques for assessment having due regard to statistical and sampling methods;
- to promote the conduct of assessment in co-operation with local education authorities and teachers;
- to identify significant differences of achievement related to circumstancces in which children learn, including the incidence of under-achievement, and to make the findings available to those concerned with resource allocation within the Department, local education authorities and schools.

The Assessment of Performance Unit has been advised from the beginning on broad outlines of policy by a widely based Consultative Committee, while a Statistics Advisory Group has offered advice on statistical matters. Steering Groups, called Working Groups initially, have been responsible for the monitoring policy in specific curriculum areas; the membership of the Steering Group on Science is given in Appendix 9.

The Science Working Group published a Consultative Paper 'Assessment of scientific development' (DES, 1977) on plans for the assessment of pupils' scientific development at about the time that the monitoring teams began work at Leeds University and at Chelsea College (amalgamated from 1985 as King's College), London University. The team based at Chelsea College has been responsible for monitoring at ages 11 and 13; the team based at Leeds University has been responsible for monitoring at age 15 and for data analysis at all three ages.

The general assessment policy adopted by APU regarded science as one of the key curriculum areas, concerned with 'particular ways of thinking about and tackling problems rather than as a label to cover particular school subjects' (DES, 1977). Such a policy required that monitoring of science should lay emphasis 'on the particular processes and skills which should be the distinctive outcomes of science education . . .' (DES, 1977). It followed that science monitoring would emphasise science processes, although it was recognised that such processes could not be divorced from the content and concepts of science, nor, indeed, from the context which specifies the circumstances in which a particular event occurs, be it everyday, in school science or in some other school subject.

As a result of an analysis of the many written responses to the Consultative Paper, further discussion within the Steering Group on Science and detailed work by the monitoring teams, the main outlines of the monitoring design and assessment framework were published in 1979 in a widely distributed 'Science Progress Report 1977–78' (DES, 1979a).

An important feature of the development of the science monitoring programme has been the involvement of the science education community. Many responses to the Consultative Paper were sent to the Assessment of Performance Unit on behalf of institutions, associations and groups during the early part of 1978; in 1979 a meeting was convened to consider the validity of the assessment framework. Several liaison groups, consisting mainly of teachers, were set up and actively involved in categorising and shredding questions in preparation for the first surveys in 1980. Members of staff at the Centre for Science and Mathematics Education, Chelsea College, and at the Centre for Studies in Science and Mathematics Education, University of Leeds, as well as science advisers and many others involved in science education contributed by writing and shredding questions and offering advice.

One of the tasks associated with the terms of reference for APU was to involve local education authorities and teachers in the conduct of assessment. The assessment of experimental skills using group practical or individual practical modes of testing has involved a large group of practising science teachers in close liaison with the monitoring teams, while other teachers have helped with the marking of responses to 'pencil and paper' tests. The monitoring programme could not have achieved its purpose without the full cooperation of the whole science education community; in particular, the conduct of the surveys depended on the goodwill of staff and pupils in the survey schools.

1.3 The assessment framework

The monitoring programme was designed to cover the full range of pupil ability at three ages (11, 13 and 15), and to reflect a view of science which encompassed both the procedural aspects of experimental work and the applications of conceptual knowledge in a variety of situations or circumstances. A framework of assessment, which is able to inform question writers and which can be used for reporting the results of the assessment, needs to reflect such a view of science.

An assessment framework, published in the 'Science Progress Report 1977–78' (DES, 1979a), envisaged the description of test questions in terms of three characteristics: scientific *process*, science *concepts* and *context*. The process list of categories and subcategories of scientific activities has undergone only minor modification and some simplification since 1980. Table 1.1 shows the categories of scientific activities applicable at ages 13 and 15. Subcategories shown in bold type correspond to those which are now regarded as major assessment subcategories in the light of monitoring experience.

Table 1.1 *Categories of scientific activities (ages 13 and 15)*

Category	Subcategories	Mode of testing
1 Use of graphical and symbolic representation	**Using graphs, tables and charts** Using scientific symbols and conventions	Written
2 Use of apparatus and measuring instruments	**Using measuring instruments** **Estimating physical quantities** **Following instructions for practical work**	Group practical
3 Observation	**Making and interpreting observations**	Group practical
4 Interpretation and application	i **Interpreting presented information** Judging the applicability of statements to data Distinguishing degrees of inference	Written
	ii **Applying: biology concepts** **physics concepts** **chemistry concepts** Generating alternative hypotheses	
5 Planning of investigations	**Planning parts of investigations** **Planning entire investigations** Identifying or proposing testable statements	Written
6 Performance of investigations	**Performing entire investigations**	Individual practical

Whereas a *process* is a specified activity intended to lead towards a result, a *concept* is a general idea derived from specific instances or occurrences–it is a way of thinking about things, whether concrete or abstract. The assessment framework is intended to minimize the concept dependence of performance on process activities. However, the importance of testing the application of science concepts in specific situations is recognised in the process activity **Applying biology, physics and chemistry concepts**. It should not be inferred that pupils with a better understanding of concepts will not be able to show this in improved performance in other categories. The characteristics of *process* and *concepts* are not entirely independent quantities – the aim of the framework design was simply to allow some measure of independence in judging performance on these important characteristics of science.

Table 1.2 shows the list of science concepts and knowledge on which application questions have been based. The list relates to the three main concept regions of Biology, Physics and Chemistry, and was agreed for all three ages by the monitoring teams and the Steering Group after extensive consultation with groups of teachers and others involved in science education (Driver *et al*, 1982, p192).

Table 1.2 *List of science concepts and knowledge*

Region	Area		
Biological			
	A	Interaction of living things with their environment	Interdependence of living things The physical and chemical environment Classification of living things Physical and chemical principles needed to interpret life phenomena
	B	Living things and their life processes	The cell Nutrition Respiration Reproduction Sensitivity and movement
Physical			
	C	Force and field	Movement and deformation Properties of matter Forces at a distance The Earth in space
	D	Transfer of energy	Work and energy Current electricity 'Waves'
Chemical			
	E	The classification and structure of matter	States of matter Pure substance Metals and non-metals Acids and bases Periodic table Atomic model
	F	Chemical interactions	Solutions Reactivity Properties of a chemical reaction Some chemical reactions

The full lists of statements which define in detail the level of complexity appropriate to the three ages of monitoring have been published (age 11: Harlen *et al*, 1981, pp213–214; age 13: Schofield *et al*, 1982, pp160–166; age 15: Driver *et al*, 1982, pp193–202). The age 15 statements are included in Appendix 7 of this report. These statements define the boundaries for question writers at each age. Because of the considerable variation in the content of the syllabuses in the different science subjects and the varied science experiences of pupils, the concept list for age 15 reflects, as far as possible, the science concepts that most pupils will have been taught by the end of their third year of secondary school.

The third characteristic of the assessment framework specifies the situations or circumstances in which the processes and concepts of science are applied to an event, object, information or data, ie the content of a question set in a particular *context*. The *contexts* were identified as science lessons – usually subdivided into Biology, Physics or Chemistry, non-science lessons and everyday or out-of-school situations.

The assessment framework, with its defined characteristics, represents a view of science which has gained increasing acceptance by the science education community during the last decade. As a result of feedback from the surveys, there has been a fruitful interaction between the development of questions and mark schemes relating to the detailed statements of process, concept and context, which has led to modifications to these detailed statements. As a consequence, the framework, the questions and the mark schemes now represent a tested and self-consistent assessment scheme, which uses a variety of question formats and modes of testing (written, circus of practical tasks and individual practical investigations) to report on the science performance of pupils. Details of the various modes of testing used are given in Chapters 5 to 11, where each science activity category is discussed in turn.

1.4 Sampling and reporting procedures

The five annual surveys of 15 year old pupils were conducted in the November of years 1980 to 1984 inclusive, and each involved a total representative sample of about 15,000 pupils, or just under two per cent of the population of 15 year olds in England, Wales and Northern Ireland. The selection of pupils was random within each school, and the schools (about 500 per survey) were themselves randomly selected to represent the range of school types, sizes, age range and location. The sample of pupils was then subdivided into parallel subsamples, each of which was given a different test package. Although each year questions equivalent to a single test lasting 20 to 30 hours were used, most individual pupils were tested for no longer than one hour, with a subsample of 'written test' pupils also taking a one hour 'practical'. This survey design, known as light sampling, was specified in the APU Consultative Paper (DES, 1977, p11) to allow the production of data on the performance in science of the population as a whole, rather than of individuals. Indeed, no pupil, school or local education authority has been, or can be, identified in the reports.

The work of selecting schools and pupils according to the two stage sampling scheme and of administering the surveys in the various curriculum areas has been co-ordinated for the APU by the Monitoring Services Unit at the National Foundation for Educational Research (NFER). It should be noted that participation by schools was on a voluntary basis and that headteachers of each selected school had discretion to withdraw pupils from the testing sessions if it was felt that taking part could cause undue distress. Otherwise the only pupils explicitly excluded from the surveys were those in special schools or in units designated as 'special' in normal schools.

The total number of questions which had passed 'shredding', validating and trialling procedures to be included in the age 15 question bank in time for the survey in 1984 was about 1,000. Some subcategory pools contain a limited number of questions from which the tests were constructed. Other subcategory pools, often referred to as domains, contain between 100–150 questions from which the questions used in the survey were chosen at random, the term used to describe this strategy being domain sampling (see Section 4.1). Further details about the development and selection of science questions at age 15 during the early stages of the project may be found in the report of the 1980 survey (Driver *et al*, 1982).

During the early years of the project's life individual questions were written with a *particular* age in mind, with a view to producing question pools as appropriate as possible for each of the three ages to be surveyed. In consequence, essentially independent pools of questions were developed, one pool per age group. More recently, in an attempt to increase the potential for performance comparisons across ages, the available questions were reviewed and 'absorbed' into other age pools where this strategy seemed appropriate. At the same time the structure of each category was reassessed with a view to increasing its contribution to the interpretation of pupil performance.

Different subcategories were affected to different degrees by this rationalization process. Least affected were the **Applying science concepts** subcategories: there are clear educational reasons why questions devised to probe the conceptual understanding of 15 year olds would not be appropriate to put before 11 year olds. The subcategories **Using graphs, tables and charts** and **Interpreting presented information** were heavily affected by large-scale absorption of questions from one age pool into another. Most radically changed by the rationalization were the subcategories **Making and interpreting observations** and **Planning parts of investigations**. These two subcategories were extensively restructured just prior to the 1983 surveys, with entirely common question pools being created for 13 and 15 year olds.

A review of the build-up and rationalization of the question bank is included for each category as appropriate in Chapters 5–11, as well as in Chapter 12, where progression in performance from age 13 to 15 is described; and in 'National assessment: the APU science approach' (Johnson, 1988).

Test results are normally presented either as a mean subcategory score or by discussion of performance on individual questions. A mean subcategory score is estimated for a specified population as a whole, based on the performance of a given subsample of pupils who took the subcategory test. Subcategory scores are most useful when considering performance in relation to variables such as 'type of school' or 'sex of pupils'. Individual question data, on the other hand, are calculated from the scores on a given question, and are particularly useful when considering specific aspects of pupils' understanding.

When reporting on performance in relation to individual pupils' abilities, the number and level of external examinations for which each sample pupil was likely to be entered at the end of the academic year are used as an indication of the pupil's general level of academic ability. Six 'examination entry' groups are defined for this purpose: 8+ O-levels, 6–7 O-levels, 3–5 O-levels, 1–2 O-levels, 4 or more CSEs, 0–3 CSEs (pupils in these last two 'CSE' groups would not be taking any O-levels). The percentage of pupils in each 'examination entry' group is both 'country' and 'boy-girl' dependent, but varies typically between approximately 10 per cent and 30 per cent of the total entry (see Table 2.8 in Chapter 2).

Because of the limited flexibility and judgemental nature of the 'examination entry' measure, it has been found useful, within most subcategories, to adopt a second measure based on scores on APU test packages. Five approximately equal sized groups have been defined on this basis: most able (top 20 per cent), above average, average (middle 20 per cent), below average, least able (bottom 20 per cent) (see for example Table 5.1 in Chapter 5).

In order to relate test results to information about the general provision for science, schools participating in the surveys were asked to complete a questionnaire which asked for information about teacher qualifications, science teaching resources, the science curriculum offered and followed, and financial provision for science. The school questionnaire used in the age 15 survey in 1984 is given in Appendix 1. Pupil questionnaires were administered for the first time in the 1984 survey, so that information could be gained about pupils' scientific interests and attitudes. These are reproduced in Appendix 2.

1.5 Structure of the report

Chapter 2 deals with science provision and subject uptake in schools – a summary of the findings from the school questionnaires over the five year period as regards resource levels, subjects offered to pupils and the pupils' option choices.

Chapter 3 considers the findings from the pupil questionnaires relating to pupils' views about their optional science subjects, the degree of gender stereotyping in their job perceptions, their assumptions about the usefulness of Biology and Physics to a range of occupations, and their awareness of, and interest in, a number of topical scientific applications.

Chapter 4 moves to a consideration of pupils' science performance and reviews performance patterns over the five year period for five domain-sampled subcategories: **Using graphs, tables and charts, Interpreting presented information, Applying biology concepts, Applying chemistry concepts** and **Applying physics concepts**.

Chapters 5 to 11 summarise the findings over the five year period for each of the science activity categories defined in the assessment framework: **Use of graphical and symbolic representation; Use of apparatus and measuring instruments; Observation; Interpretation; Application of science concepts; Planning investigations** and **Performing investigations**.

Chapter 12 summarises the findings on age related patterns of performance with particular reference to progression from age 13 to 15.

Chapter 13 focuses on curriculum influence by exploring connections between subcategory performance and option choices in science.

Chapter 14 assesses the overall results of the five surveys at age 15 and considers some issues raised by the whole exercise for science education.

Comments are occasionally made about the 'significance' of particular performance findings, in particular where these relate to comparisons between pupil or school groups, such as boys and girls or inner city versus prosperous suburban schools. Whenever the term 'significance' is used this refers to *statistical* significance at the *five per cent level*. It should be recognised that statistical significance is not synonymous with educational significance. With large samples rather small differences often reach statistical significance when in practice they are of little educational importance. Similarly it does not follow from a lack of statistical significance that a performance difference should be ignored. Differences which persistently emerge in the same direction in most or all surveys are likely to be worthy of attention and may well have pedagogical implications, whether or not they reach statistical significance in any one survey.

2

Science provision and subject take-up

2.1 Introduction

Option choices at 13+ result in a wide variety of curriculum experiences in science among 15 year olds. This has major implications for the interpretation of the results of science performance surveys of the kind considered in this report. After all, it is one thing for a 15 year old to be unable to interpret a solubility graph when that pupil discontinued any study of Chemistry at 13+. It is quite another thing when an O-level Chemistry student fails in this kind of task. From the earliest days of the science assessment programme it was recognised that information about the science performances of 15 year olds would be much impoverished without supplementary knowledge about subject take-up.

Similar arguments can be made for the need to gather information about the resources which schools have available for science, and about the range of courses – or, at primary level, variety of science relevant experience – they provide for their pupils.

In every APU science survey to date questionnaire information has been gathered from the participating schools so that this kind of information might become available. For example, schools have been asked about their financial income and expenditure, about their laboratory availability and technician support, and about the science qualifications of their teachers. Additional enquiries, unique to particular ages, have been included from time to time. An example is the enquiry described later, in which schools were asked to indicate to which examination levels they provided various subject courses in science.

The participating schools in the surveys of 15 year olds have always been asked to indicate the science subjects, if any, being studied by each of their sample pupils. As mentioned in Chapter 1, they have also been asked to give an estimate of the number and level of external examinations for which each of these pupils was likely to be entered, to provide a rough indication of each pupil's general level of academic ability.

The questionnaire findings have been discussed in varying degrees of detail in the earlier survey reports at the appropriate ages. Information about subject provision and take-up at age 15 over the period 1980–82 was also reviewed and discussed in Driver *et al* (1984a). This chapter describes very briefly the survey findings on teacher qualifications and laboratory provision, and looks in some depth at the findings about subject provision and take-up. For comment about links with pupil performance readers are referred to Chapters 4 and 13.

It soon became clear that it was not possible to report meaningfully the results of the enquiries on finance. This is so for two main reasons. Firstly, too many schools either could not provide the relevant data, or chose not to do so. This problem was most acute within the Northern Ireland samples. Secondly, where information about expenditure on science *was* given a number of schools indicated that the particular figures were unusual in some respect; including, for instance, one-off grants for laboratory refurbishment, new course introduction, capital equipment purchase, and so on. Information about schools' general levels of capitation income is published regularly elsewhere; in particular, in the annual publications of the Chartered Institute of Public Finance Accountants (CIPFA).

2.2 Teacher qualifications and laboratory availability

The information about laboratory availability and teacher qualifications which has accumulated over the five year survey period has been fully discussed in the age 13 review report (see Schofield *et al*, 1988). A brief overview only will be given here.

On teacher qualifications the salient findings are that, in each country, about 95 per cent of the teachers of science in the survey schools held an academic qualification in a science or related subject. Almost three-quarters of these teachers were science graduates, most others holding BEd degrees with science a main subject. There was a marked difference between England and Wales on the one hand and Northern Ireland on the other in terms of the proportions of science graduates who held a Postgraduate Certificate in Education: two in three of the English and Welsh science graduates compared with just one in three of those in Northern Ireland.

In all three countries Biology was the most common subject of qualification (about a third of all science teachers), followed by Chemistry (about one in four) and Physics (about one in five). Among the teachers of science to 13 year olds there were even lower proportions of teachers with qualifications in Chemistry (about one in five) and Physics (roughly one in ten). This suggests that the physicists and chemists among the teachers with specialist qualifications are rather fully employed with examination classes in the fourth and fifth years and in the sixth form.

As far as laboratory availability is concerned the evidence from the surveys is that, in England and Wales at least, most schools have enough laboratories to meet their *present* needs (which might, of course, have themselves been dictated by the accommodation available). There was no indication that 13 year olds were being deprived of laboratory time because of pressure higher up the school: only a handful of science classes at this age were receiving all their science lessons in a classroom.

However, it was also clear that in general schools were not *over supplied* with laboratories, so that were extra demands on laboratory space to be introduced, then the schools' present laboratory allocation practices would have to be modified to accommodate these. For instance, the evidence is that at least a third of all comprehensive schools in England and Wales would currently be unable to provide fully *practically based* science courses for *all* their pupils aged between 11 and 16 of the kind currently advocated (see DES, 1985). Grammar and Independent schools *are* sufficiently well equipped with laboratories for this kind of change to be possible.

Of all the schools which have taken part in the science surveys the girls' schools in Northern Ireland have been by far the least well resourced in terms of laboratory accommodation. Among *these* schools the Secondary Intermediates fared least well. The evidence is that few of the girls' schools in Northern Ireland have enough laboratories even to think about reducing the persisting gap in the proportions of boys and girls studying any science at all after 13+. They certainly have insufficient laboratory accommodation to expand their present levels of provision of *practically based* science education.

2.3 Subject provision and school policies for science take-up

In the questionnaires from 1981 onwards the schools which participated in the surveys of 13 or of 15 year olds were asked to indicate the range of science subject courses they were providing for their pupils. Specifically, they were presented with a list of named subjects and were asked to mark these appropriately.

Before discussing the findings from this enquiry, it is of interest to note that, in the lower school, the evidence is that almost three-quarters of Comprehensive schools provide a single combined science course for all their third year pupils. About another tenth constrain all their 13 year olds to follow separate subject courses in the three main sciences; this was the most common practice in Grammar and Independent schools. Among those schools not following either of these strategies, the most frequent practice is for the more academically able pupils to follow separate subject courses and for the average and less academically able pupils to study an integrated course of some kind (see Schofield *et al*, 1988, for details).

Sometime during their third year in secondary school, pupils are normally required to make important decisions about the subjects they will choose to study for their next two years in school – and perhaps beyond. It is likely that among the many influences on these option choices will be the *range* of courses offered to them by their schools.

Table 2.1 *Science subject provision in the survey schools in 1984*

(Percentage of schools providing O-level, CSE Mode 1 or Joint GCE O-level/CSE courses in the indicated subject)

Subject	England	Wales	Northern Ireland
Biology*	88	97	84
Chemistry	89	95	70
Physics	91	97	84
General Science	33	46	36
Human Biology	37	53	30
Rural Science	8	27	0
Electronics	8	8	0
Technology	6	5	4
Environmental Science	3	0	6
Physics-with-Chemistry	4	1	0
Physical Science	4	4	0
Number of schools	263	100	97

* Biology courses in Wales were exclusively Joint GCE O-level/ CSE examination courses at the time of testing.

Table 2.1 gives the percentages of survey schools which were providing O-level, CSE Mode 1 or joint O-level/ CSE courses in each named subject in 1984. As the table shows, courses in the three main sciences were provided in more than 90 per cent of the survey schools in England and Wales. In Northern Ireland, on the other hand, all three of these subjects were relatively less widely available, with Chemistry the least frequently represented. This difference in provision is most likely a consequence of the differences in the school systems in these countries: while the state systems in England and Wales are now almost totally Comprehensive, that in Northern Ireland is still almost entirely selective. The schools in Northern Ireland *not* providing courses in the main sciences are all Secondary Intermediate schools.

Nuffield courses were provided by fewer than one in ten schools in England and Wales, and were not available at all in any of the survey schools in Northern Ireland. On the other hand, *many* of the survey schools made some – in many cases substantial – use of the Nuffield course materials.

Another feature of note in Table 2.1 is the relative popularity of General Science, Rural Science and Human Biology in the schools in Wales.

Table 2.1 is concerned with O-level, Mode 1 CSE and joint GCE/CSE courses, but many schools also offered Mode 3 CSE or non-examination courses in these same subjects. About a third of the survey schools in England and Wales and just one in twenty of those in Northern Ireland provided courses at this level in at least one science subject. Few of these were Grammar or Independent schools.

The most frequent combinations of subjects provided in individual schools included the three main sciences, as Table 2.2 shows. About a fifth of the schools in England and Wales and 27 per cent of those in Northern Ireland (the Grammar schools) provided courses in the three traditional specialist sciences, but in no others. Another 20–40 per cent of the schools in each country provided Human Biology or General Science or both in addition. The remaining schools provided various different combinations ranging from a single science subject (General Science), through two or three subjects (such as Biology and Physics, or Biology, Physics and General Science) to as many as six different subjects (such as Biology, Chemistry, Physics, General Science, Human Biology and Electronics).

Table 2.2 *Combinations of subjects provided by the schools*

(Percentage of schools making courses available in all the named subjects)

Subjects	England	Wales	N Ireland
Biology, Chemistry and Physics	19	18	27
Biology, Chemistry, Physics, Human Biology	13	17	6
Biology, Chemistry, Physics, General Science	16	9	11
Biology, Chemistry, Physics, Human Biology, General Science	5	15	6
Other subject combinations (1 to 6 subjects)	47	41	50
Number of schools	263	100	97

This then is the picture in terms of subject provision. But what do we know about the choice constraints, if any, which schools place on their pupils? Indeed, what freedom do pupils have to avoid science altogether at 13+? To attempt an answer to this question, the schools in the 1984 survey were asked whether any constriants were imposed on their pupils of *average* ability, ie on those pupils taking at least two O-levels or mainly CSEs. The responses to these questions are given in Table 2.3. The table shows immediately that about 80 per cent of the English schools, about two-thirds of those in Wales and just about a third of those in Northern Ireland constrain their pupils of generally average academic ability to take at least one science subject in their fourth and fifth years.

Table 2.3 *Schools' policies for science take-up by average ability pupils at 13+*

(Percentage of schools indicating each policy)

At least one *science* subject to be taken by those pupils:	England	Wales	Northern Ireland
—entered for at least 2 O-levels	82	66	31
—taking mainly CSEs	79	66	35

2.4 Option choices

As mentioned earlier, in every science survey at this age the schools are asked to indicate the science subjects being studied by each sample pupil. A standard list of subjects with associated codes has been provided for this purpose: Biology, Chemistry, Physics, Physics with Chemistry, Human Biology, General/Combined/Integrated Science, Rural Science, Electronics, Technology, SCISP, 'other science' and 'no science'. The effect of the schools' policies on science take-up in the three countries is readily seen in Table 2.4, in terms of the *amount* of science being studied at this age by the pupils.

Table 2.4 *Numbers of science subjects studied by 15 year olds in 1984*

(Percentage of 15 year olds studying indicated numbers of subjects)

Number of subjects	England			Wales			Northern Ireland		
	Boys	Girls	All	Boys	Girls	All	Boys	Girls	All
Three or four*	15	8	12	16	13	14	7	10	11
Two	34	25	30	29	23	26	18	15	16
One	48	60	54	48	55	51	53	46	50
None	3	6	4	7	9	8	21	28	25
No. of pupils	3600	3547	7147	1378	1286	2664	1298	1512	2810

* Fewer than 1 per cent of all survey pupils were studying four science subjects.

Table 2.4 shows a generally greater take-up of science by pupils in England and Wales compared with Northern Ireland. Very small percentages of English and Welsh pupils drop science altogether, whereas a relatively high proportion of 15 year olds in Northern Ireland schools are still able to do so. The majority of pupils in all three countries study just one science subject. As will be shown later, these are mainly the average and less able pupils; they will typically be following a combined science course of some kind, or, if boys, Physics or Rural Science, or, if girls, Biology or Human Biology. The majority of those pupils studying *three* subjects will be the most able pupils (8+ O-levels); the subjects will be Biology, Chemistry and Physics, all taken to O-level.

Another feature readily seen in the take-up data in Table 2.4 is the differential amount of science studied at this age by boys and girls. In every survey the take-up enquiry has shown proportionally more girls than boys among those pupils in the 'no science' group. In Northern Ireland, in particular, almost 30 per cent of the girls tested in the 1984 survey had studied no science at all since they made their option choices. This is certainly in line with the findings about laboratory availability in Northern Ireland schools. As mentioned earlier, the girls' schools in Northern Ireland – Grammar and, particularly, Secondary Intermediate – were much less well provided with science laboratories than were the boys' schools in the survey samples. It would be difficult to increase the numbers of girls studying science in this country without creating rather serious resource problems.

In all three countries the evidence is that take-up of science in some form or another has increased steadily over the five year period of the surveys, among boys *and* girls in England but mainly among girls in Wales and Northern Ireland. Table 2.5 illustrates this feature in the survey data, showing the percentages of 15 year olds in each survey sample who were studying *no* science at all.

Table 2.5 *'Drop-out' from science at 13+ over the period 1979–83**

(Percentage of 15 year olds in each survey sample 1980–84 with no science subject among their option choices)

	England			Wales			Northern Ireland		
Year	Boys	Girls	All	Boys	Girls	All	Boys	Girls	All
1980	5	11	8	8	14	11	21	38	29
1981	4	7	6	8	12	10	16	33	25
1982	4	9	6	5	11	10	18	31	24
1983	4	7	5	7	9	8	18	27	23
1984	3	6	4	7	9	8	21	28	25

* As with all the take-up data, these figures are unweighted sample statistics. Sample sizes vary between 1,600 and 2,800 pupils in Wales and in Northern Ireland over the period, and between 6,500 and 11,500 in England.

Indeed, as noted in an earlier paper (Driver *et al*, 1984b), there are indications that take-up had already increased in England over the five year period before the APU surveys began (no comparative information is available for Wales or Northern Ireland). Table 2.6 presents the relevant figures, including as it does comparative figures derived from the report of HMI's Secondary Survey (DES, 1979b). The increase in take-up can usefully be averaged over the entire 10 year period spanned by the HMI and the APU surveys. The result is that each year one per cent more of the population of 15 year olds were studying at least one science subject.

Table 2.6 does suggest that there has been an increase over the past 10 years in the amount of science now studied by 15 year olds. But has there been any corresponding change in the *pattern* of individual subject take-up? A review of the take-up rates for particular subjects, described later, shows indeed that there *has* been a change in the *relative* take-up rates of different science subjects. Much of the overall increase has been to the benefit of Chemistry and Physics – a feature which has been noted also in school leaver statistics (see Tall, 1985). And, as we shall see later, it is the *girls* more than the boys who are taking these subjects in increasing numbers.

Table 2.6 *Science take-up in England since the mid-70s*

(Percentage of 15 year olds studying indicated numbers of subjects)

	1975/8*			1980			1984		
Number of subjects	Boys	Girls	All	Boys	Girls	All	Boys	Girls	All
Three or four	10	4	7	16	8	12	15	8	12
Two	31	18	25	32	22	27	34	25	30
One	50	60	55	47	59	53	48	60	54
None	9	18	13	5	11	8	3	6	4
Pupils (1000s)	19.2	16.9	36.1	5.8	5.9	11.8	3.6	3.6	7.2

* The earlier figures have been abstracted from Table 8A of the HMI Secondary Survey Report (DES, 1979).

Table 2.7 provides details about the proportions of survey pupils who were studying various science subjects in 1984. The predominance of the traditional subjects in the pupils' option choices is clear, reflecting to some extent the wide availability of these subjects in the schools. As in past surveys very few pupils have been found to be studying the less well established subjects. In particular, Electronics and Technology are still studied by extremely small proportions of pupils. The 'other subject' group shown in Table 2.7 will include a very wide range of less frequently provided science and science related subjects, each taken by extremely small numbers of pupils in even smaller numbers of schools. Examples given in the schools' responses are Engineering Physics, Marine Biology, Human and Social Biology, 'Science at work', Agricultural Science, Geology, and CDT.

Table 2.7 *Subject take-up rates in 1983*

(Percentage of 15 year olds studying indicated subjects in 1984 – alone or with other sciences)

Subject	England	Wales	Northern Ireland
Biology	45	40	35
Chemistry	32	29	22
Physics	38	38	30
General Science	12	15	12
Human Biology	8	9	4
Rural Science	3	5	3
Technology	2	<2	1
Electronics	<1	<1	0
Physics with Chemistry	<2	<2	<1
SCISP	<2	0	0
Other science subject	6	6	3

Take-up rates are highest in England for the three main sciences, and lowest in Northern Ireland. For General Science, Human Biology and Rural Science, Wales shows the highest levels of take-up–though by slight amounts. Again, this reflects the wider availability of these particular subjects in the survey schools in Wales (see Table 2.1).

About a fifth of the pupils were experiencing a 'balanced' science curriculum, taking the form either of all three specialist science subjects (about 10 per cent of pupils), or of a single combined science course (about 10 per cent).

In previous survey reports it has been noted that courses in these various subjects were not taken equally by pupils of different abilities. It is well known that schools do tend to 'channel' pupils of different ability levels into particular kinds of science course. Specifically, only the most able pupils usually follow courses in all three main sciences. Moreover, while pupils of average ability and below often follow courses in Biology only or in Physics only, this is unlikely in the case of Chemistry. Typically, average and less able pupils take a General Science course, or, alternatively, study Human Biology (if girls) or Rural Science (if boys). Indeed, this general division of pupils of different abilities into the two main kinds of science curriculum is often implemented in the third year or sooner, as the surveys of 13 year olds have shown (see, for instance, Schofield *et al.* 1988). The outcomes of this 'channelling' are considered in the next section.

2.5 Ability-related differences in subject take-up rates

As mentioned earlier, the survey schools are asked to indicate both the number and level of external examinations each sample pupil is likely to be entered for the following year, ie towards the end of their fifth year. This information has always served as a useful indicator of the pupils' general levels of academic ability. Before presenting the take-up figures broken down by ability, it will be useful to look first at the total make-up of the pupil sample in terms of their examination entry prospects. This information is shown in Table 2.8.

The most striking feature in Table 2.8 is the difference in the pattern of examination entry intentions shown for the Northern Ireland sample compared with those for England and Wales. The Northern Ireland distribution is much less uniform than the others, with a particularly high proporiton (29 per cent) of pupils to be entered for eight or more O-levels and a rather small proportion (seven per cent) to be entered for just one or two O-levels. This difference is notable in the examination entry data for all five annual science surveys. It will be a consequence of the still selective school system of Northern Ireland: Grammar schools tend to enter pupils for O-levels in the main, while Secondary Intermediate schools rarely enter pupils for this level of examination. This feature in the national examination entry data serves as a reminder that examination entry intentions reflect school policies to a great extent, and can therefore be considered as only a crude indicator of pupil ability levels. They have, nevertheless, proved useful in the survey context in some respects.

Another point to note in Table 2.8 is the tendency for girls to have better examination prospects than boys in all three countries. Again, the Northern Ireland sample is unique in the quite severe discrepancy in the proportions of boys and girls who were to be entered for eight or more O-levels, and in the proportions who were to take just a handful of CSEs or no examinations at all.

Table 2.9 (p11) shows the subject take-up rates among pupils of different ability levels in the three countries, for those subjects taken by at least three per cent of the entire sample from England. The table illustrates clearly the result of the schools' policies of 'channelling' their more able pupils into courses in the main sciences, and their less able pupils into courses in Human Biology, Rural Science or some form of Combined Science.

Roughly 60 per cent of the most able pupils (8+ O-levels) in all three countries were following courses in Biology, Chemistry or Physics. A third of the most able pupils were indeed studying all three main subjects – these would be the 'Grammar' school pupils in the old selective system (they *are* the Grammar school pupils in Northern Ireland). In contrast, fewer than one per cent of the pupils in this ability group were following courses in Human Biology, Rural Science or General Science.

Table 2.8 *Pupil-sample composition in 1984 in terms of examination entry plans*
(Percentage of pupils in each 'examination entry' group)

Examination intentions*	England			Wales			Northern Ireland		
	Boys	Girls	All	Boys	Girls	All	Boys	Girls	All
8+ O-levels	21	19	20	21	24	23	26	32	29
6–7 O-levels	11	13	12	13	15	14	6	13	10
3–5 O-levels	15	17	16	17	16	16	11	10	11
1–2 O-levels	13	15	14	14	13	14	7	7	7
4 or more CSEs	26	25	25	20	19	19	24	24	24
0–3 CSEs	15	12	13	15	13	14	26	15	20
Number of pupils	3378	3297	6675	1292	1222	2514	1283	1483	2766

* Those pupils in the first four 'O-level' groups might also be taking CSEs; those pupils in the two 'CSE' groups would *not* be taking O-levels.

Table 2.9 *Subject take-up rates among survey pupils of different abilities in 1984*

(Percentage of pupils in each ability group taking indicated subject)

Subject	Country	O-levels 8+	6–7	3–5	1–2	CSEs* 4+	0–3
Biology	England	66	59	56	46	34	17
	Wales	70	60	47	31	17	5
	Northern Ireland	63	52	33	25	23	5
Chemistry	England	62	49	36	28	15	6
	Wales	58	42	32	18	11	2
	Northern Ireland	60	22	10	4	4	<1
Physics	England	62	52	46	34	26	11
	Wales	60	50	46	32	22	6
	Northern Ireland	60	28	24	23	20	6
General Science	England	1	3	5	8	18	36
	Wales	1	3	7	14	31	36
	Northern Ireland	1	5	7	10	15	28
Human Biology	England	4	5	5	9	11	11
	Wales	<1	3	9	16	17	11
	Northern Ireland	1	3	7	9	7	3
Rural Science	England	<1	<1	1	3	5	4
	Wales	<1	0	3	8	11	8
	Northern Ireland	0	<1	0	<1	2	10

* These 'CSE pupils' were not taking any O-levels.

There are some interesting national differences in Table 2.9. In particular, the data show a higher take-up of General Science, Human Biology and Rural Science among the average and less able pupils in Wales compared with their contemporaries in England and Northern Ireland. These findings reflect the higher availability of these subjects in the schools in Wales.

While Physics take-up rates were similar across the ability range in England and Wales, take-up of Biology and of Chemistry tended to be higher in England, particularly among average and less able pupils. For all main science subjects take-up rates were lowest among Northern Ireland pupils for all but the most able pupils; markedly so for both Chemistry and Physics.

Table 2.10 *The extent to which pupils of different abilities follow a 'balanced' science curriculum at age 15*

(Percentage of pupils in each ability band following indicated courses)

Subjects	Country	O-levels 8+	6–7	3–5	1–2	CSEs* 4+	0–3
Physics, Chemistry and Biology	England	33	16	7	2	1	<1
	Wales	31	13	13	2	2	0
	Northern Ireland	29	2	<1	1	<1	0
General Science	England	<1	1	4	7	14	31
	Wales	<1	3	6	8	20	27
	Northern Ireland	1	3	7	10	12	25

* These 'CSE pupils' were not taking any O-levels.

Table 2.10 gives the percentages of pupils in the different ability bands who were following a 'balanced' curriculum in science. As noted earlier, about a fifth of all the sample pupils were experiencing a science curriculum containing elements of all three main sciences, either in the form of separate subject courses or as a single General Science course. Table 2.10 shows this degree of experience to be very uneven across the ability range. About a third of the most able pupils and a similar proportion of the least able pupils were benefiting from a balanced curriculum, compared with as few as 10 per cent of the average ability pupils. The pupils in the most able group were in the main following separate subject courses in the three main sciences–generally all to O-level, while those in the least able group were studying some form of General Science.

Given the different subject take-up rates among the groups of pupils at different ability levels, it is not surprising that the ability compositions of the samples of pupils taking particular subjects vary enormously. The majority of those pupils studying the main sciences were the most able pupils; the majority of those studying Human Biology, Rural Science or General Science were the least able pupils. The ability distributions among particular subject takers also vary appreciably from one country to another, reflecting the differential patterns of ability related take-up rates shown in Table 2.9. Table 2.11 illustrates these findings.

Table 2.11 *Ability distributions among those survey pupils studying particular subjects in 1984*

(Percentage of pupils of indicated ability level among those taking subject)

Subjects	Country	O-levels 8+	6–7	3–5	1–2	CSEs* 4+	0–3
Biology	England	28	15	19	14	19	5
	Wales	39	21	19	11	8	2
	Northern Ireland	52	15	10	5	15	3
Chemistry	England	38	18	18	12	12	2
	Wales	45	20	18	9	7	1
	Northern Ireland	80	10	5	1	4	<1
Physics	England	32	16	19	12	17	4
	Wales	36	19	20	12	11	2
	Northern Ireland	58	9	9	5	16	4
General Science	England	2	2	7	10	38	41
	Wales	2	2	8	14	41	33
	Northern Ireland	3	4	6	6	30	50
Human Biology	England	10	8	10	16	37	19
	Wales	1	5	16	24	37	16
	Northern Ireland	7	7	18	14	41	14
Rural Science	England	2	1	5	15	52	24
	Wales	2	0	10	22	44	22
	Northern Ireland	0	3	0	1	15	81

* These 'CSE pupils' were not taking any O-levels.

Table 2.11 shows that in the case of Biology, Chemistry and Physics the ability distributions are more skewed for Wales than for England, and are *most* severely skewed for Northern Ireland. The make-up of the Chemistry sample in Northern Ireland is especially noteworthy, with fully 80 per cent of the Chemistry takers coming from the *most able* group of pupils. These Northern Ireland differences will, of course, appear exaggerated, given the relatively greater proportions of pupils classified as 'most able' in this country.

In the case of General Science, Human Biology and Rural Science the distributions for England are only slightly more skewed than are those for Wales. For General Science and Rural Science the Northern Ireland distributions are again the most severely skewed. Indeed, Rural Science parallels Chemistry in Northern Ireland, with fully 81 per cent of the Rural Science takers being drawn from one particular ability group – this time the *least able* pupils.

Interestingly, the low numbers of pupils (predominantly boys) who were studying Technology and/or Electronics were drawn fairly evenly from all the ability groups (although this might not be the case in individual schools). One outcome of this is that the ability distributions for the pupil samples taking these subjects are almost uniform.

2.6 Gender differences in subject take-up

It is clear that the picture as regards science take-up in the three countries is highly complex. Particular subjects tend to be taken most often by pupils of higher or lower ability, and the strength of this tendency varies from one subject to another and from one country to another for the same subject. Indeed, this is not the complete story, since the picture is further complicated by gender differences in subject take-up. Table 2.12 begins to unfold this added complexity, by showing the subject take-up rates for boys and girls for the subjects previously listed in Table 2.7.

Table 2.12 shows that none of the named science subjects attracted the *same* proportions of boys and girls. The subject coming closest to equal 'popularity' is General Science – and even here the proportions of boys and girls following combined science courses were similar only in England and Wales. The familiar polarization of boys and girls into, respectively, the physical sciences and the biological sciences is ever present. In all three countries, proportionally more girls than boys were following courses in Biology or Human Biology. With the exception of Chemistry in Northern Ireland, the situation is reversed for the physical sciences and technological subjects, with proportionally more boys than girls following relevant courses.

Table 2.12 *Subject take-up rates among boys and girls in 1983*

(Percentage of 15 year olds studying indicated subjects in 1984)

	England		Wales		Northern Ireland	
Subject	Boys	Girls	Boys	Girls	Boys	Girls
Biology	34	57	32	49	21	46
Chemistry	36	28	31	27	20	23
Physics	56	21	50	25	43	19
Human Biology	4	11	4	14	1	7
General Science	12	11	15	14	18	6
Rural Science	4	1	4	3	4	1
Technology	3	<1	<4	0	2	0
Electronics	<2	<1	<2	0	0	0
Physics with Chemistry	2	<1	<2	<2	0	<1
SCISP	3	<1	0	0	0	0
Number of pupils	3600	3547	1378	1286	1298	1512

It was noted earlier that much of the increased take-up of science at 13+ since 1980 could be attributed to an increase in the take-up of Chemistry and Physics among girls. Table 2.13 presents the appropriate findings from the annual APU Science surveys. For ease of presentation, figures are given for England only; very similar patterns are evident in the survey data for Wales and for Northern Ireland.

Table 2.13 *Take-up of main sciences over the period 1979–83*

(Percentage of 15 year olds studying each subject in survey samples 1980–84 in England)

	No of pupils		Biology		Chemistry		Physics	
Year	Boys	Girls	Boys	Girls	Boys	Girls	Boys	Girls
1980	5793	5891	32	54	35	25	58	17
1981	3294	3143	33	55	40	26	61	19
1982	4064	4102	32	54	35	29	57	21
1983	2936	3414	32	57	37	30	58	22
1984	3600	3547	34	57	36	28	56	20

There is *some* evidence also that General Science may have declined in popularity among boys only in England and Wales, and that Human Biology is less frequently taken now by girls in all three countries.

It is difficult to comment sensibly about changes in popularity for the much less widely available subjects – such as Technology, SCISP, and Physics with Chemistry – since these are taken by extremely small percentages of pupils drawn from those rather few schools which provide relevant courses. Although it *has* consistently been the case that a mere handful of girls have been found to be studying Technology in England and Wales (typically 0.1–0.2 per cent of those surveyed – literally five or six girls at most in samples of 3,000 to 6,000 in England, one or two girls in samples of 850 to 1,250 in Wales), and none at all in Northern Ireland. Even among boys the numbers taking this subject remain low. The number of boys studying Technology has hovered around three and a half per cent of those surveyed in England and Wales, and has fluctuated between one per cent and six per cent in Northern Ireland.

For Electronics the indications are that between one and one and a half per cent of 15 year old boys and a handful only of girls in England and Wales were studying this subject over the survey period, with none at all doing so in Northern Ireland (other than a handful of boys from one school in 1981).

The association between subjects studied and general academic ability, illustrated earlier for all pupils in

Tables 2.9 and 2.10, is present among both boys *and* girls, but to differing degrees. Table 2.14 presents the relevant figures for the three main sciences. Perhaps of especial interest in Table 2.14 is a comparison of the take-up figures for these subjects among the most able pupils in the three countries.

As mentioned earlier, the school system in Northern Ireland is still selective; moreover, many of the schools in this country remain single sex schools. In contrast, the maintained school systems in England and Wales are now almost entirely comprehensive *and* coeducational. To some extent, therefore, the take-up patterns in Table 2.14 must reflect the influence of these differences in school systems. Yet the figures do not support the belief held by many science educators and others that coeducation reinforces the traditional polarization of boys and girls into the physical and biological sciences, respectively. On the contrary, the evidence is that similar or greater proportions of the boys *and* the girls take these sciences at *all* ability levels in England and Wales compared with Northern Ireland.

Table 2.14 *Subject take-up rates among boys and girls of different abilities*

(Percentage of 15 year olds studying indicated subjects in each ability group in 1984)

			O-levels				CSEs*	
Subject	Sex	Country	8+	6–7	3–5	1–2	4+	0–3
Biology	Boys	England	58	48	41	29	24	11
		Wales	61	48	43	27	9	3
		Northern Ireland	44	32	27	17	14	3
	Girls	England	74	69	70	60	44	22
		Wales	78	72	52	36	26	8
		Northern Ireland	76	61	39	33	30	7
Chemistry	Boys	England	70	57	41	31	18	6
		Wales	62	45	36	23	12	4
		Northern Ireland	61	31	9	6	4	<1
	Girls	England	53	41	31	24	13	4
		Wales	55	38	28	13	9	<1
		Northern Ireland	60	18	11	2	4	1
Physics	Boys	England	75	78	73	59	43	16
		Wales	76	70	61	49	32	11
		Northern Ireland	80	68	42	44	37	8
	Girls	England	49	31	22	12	7	4
		Wales	46	32	30	13	11	<1
		Northern Ireland	47	12	7	6	6	3

* Sample sizes as given in Table 2.9, divided evenly between the sexes.

Comprehensive reorganisation combined with an associated move towards more coeducation in England and Wales does not seem to have had a detrimental effect on the science education of pupils, at least not as regards subject availability and take-up. Indeed, the indications are that this change has in practice widened pupils' educational choices as far as science is concerned.

One outcome of these differential take-up rates among boys and girls of different abilities is shown in Table 2.15, in terms of the ability distributions among those pupils studying the main sciences. There are a number of important features in this data. Firstly, while the representation of more able boys was higher than that of more able girls in the Biology samples in England, this was not the case for Wales or Northern Ireland, where the ability distributions for this subject were very similar. In the case of Chemistry, the table again shows very similar ability distributions among the boys and the girls in Northern Ireland, with only a slight relative over-representation of the most able boys in England and a larger over-representation of the most able girls in Wales.

Table 2.15 *Ability distributions among the boys and girls studying the main sciences*

(Percentage of pupils of different abilities among those studying indicated subject)

			O-levels				CSEs*	
Subject	Country		8+	6–7	3–5	1–2	4+	0–3
Biology	England	Boys	34	15	18	10	18	5
		Girls	25	16	21	16	19	4
	Wales	Boys	39	20	22	12	6	1
		Girls	39	22	17	10	10	2
	Northern Ireland	Boys	52	9	14	5	16	4
		Girls	52	17	9	5	15	2
Chemistry	England	Boys	40	17	17	11	13	3
		Girls	36	19	18	13	12	2
	Wales	Boys	41	19	20	10	8	2
		Girls	50	21	16	6	6	1
	Northern Ireland	Boys	79	10	5	2	4	<1
		Girls	80	10	5	<1	4	<1
Physics	England	Boys	28	15	20	13	20	5
		Girls	45	19	17	9	8	2
	Wales	Boys	31	19	20	14	13	3
		Girls	45	20	19	7	8	<1
	Northern Ireland	Boys	48	10	11	7	20	5
		Girls	77	8	4	2	8	2

* These 'CSE pupils' were not taking any O-levels.

For Physics the ability distributions for all three countries show large differences in the representation of the most able boys and girls in the subject samples. The proportions of most able boys and girls among the Physics takers were roughly 30 per cent and 45 per cent, respectively, in England and Wales. In Northern Ireland the discrepancy is much greater with 48 per cent of the boys studying Physics being from the most able group compared with 77 per cent of the girls.

There are differences too among the less 'popular' subjects. Among those mainly average and less able pupils studying General Science, for instance, the representation of least able (including non-examination) pupils is greater in the samples of boys than in those of girls in England and Northern Ireland (roughly 35 per cent of the girls drawn from the least able group compared with 46 per cent and 56 per cent, respectively, of the boys). In Wales, on the other hand, the ability distributions among the boys and the girls studying this subject were similar, about a third of these pupils were drawn from the least able group.

2.7 Summary

The evidence of the surveys is that most secondary schools in England and Wales have sufficient numbers of laboratories to meet their *current* needs without constraining any pupils – even those lower down the school – to receive their science lessons in classrooms. However, *few* schools other than Independent or Grammar schools are so well provided with laboratories that they could expand their practically based science provision; for instance to enable *all* their pupils in the 11–16 age range to study science for a fifth of their curriculum time. Indeed, the situation among the girls' schools in Northern Ireland is such that they would experience great difficulty in matching the *present* levels of science provision for girls which are the norm in England and Wales.

Three-quarters of the teachers of science in the survey secondary schools were science graduates, most of the rest held BEd degrees with science a main component. Around three-quarters of these teachers held a subject qualification in one or other of the three main sciences, Biology being the most common.

The three main sciences – Biology, Chemistry and Physics – also continue to dominate the scene as far as science subject provision and take-up is concerned. The surveys have confirmed that these subjects are very widely available in schools, and they predominate among pupils' option choices. In contrast, less traditional subjects, such as Electronics and Technology, are provided by fewer than one in ten schools and, in consequence, are studied by rather few 15 year olds.

There were some notable national differences in subject provision. For instance, the three main sciences were each provided to O-level and/or CSE by 90 per cent or more of the schools in England and Wales (Biology in Wales was exclusively provided as Joint GCE/CSE examination courses at the time of testing.). In Northern Ireland availability was lower: Biology and Physics were provided in 84 per cent of the survey schools in this country, and Chemistry in a lower 70 per cent. Courses in General Science, Rural Science and Human Biology were made available by proportionally more of the survey schools in Wales than those in the other countries.

Electronics was provided by just under one in ten schools in England and Wales and in *none* of those in Northern Ireland. Nuffield courses, SCISP, Physics with Chemistry and Physical Sciences were made available in fewer than one in twenty of the survey schools in England and Wales, and again in *none* of those in Northern Ireland. Technology courses were being provided in about one in twenty sample schools in all three countries.

Schools' policies about science take-up also differed in general from one country to another. About 80 per cent of the schools in England, 66 per cent of those in Wales and just 33 per cent of those in Northern Ireland constrained their average and more able pupils to include at least one science subject among their option choices.

The evidence in the survey data is that over the period of the surveys there has been an increase in the percentage of 15 year olds studying some science. In 1984 about three to four per cent more 15 year olds were studying some science compared with the situation in 1980. The HMI Secondary Survey conducted in England in the mid-70s provides further comparative data for this country, and suggests that over the entire 10 year period to 1984 the percentage of 15 year olds studying at least one science subject has increased by one percentage point per year. The increase has been among girls only in Wales and Northern Ireland, but among *all* pupils in England. Among girls, take-up of Physics and Chemistry has increased noticeably.

Among the 15 year olds in the 1984 survey sample, the proportions studying no science at all were four per cent, eight per cent and 25 per cent in England, Wales and Northern Ireland, respectively. There were twice as many girls as boys among these pupils in England, almost equal representations in Wales, and one and a half times as many girls as boys in Northern Ireland (these ratios *reduced* from 2:1 in both Wales *and* Northern Ireland since 1980).

Subject take-up was highest for the main sciences in all three countries, and was higher in England and Wales than in Northern Ireland. Forty to forty-five per cent of the 15 year olds from England and Wales were studying Biology and/or Physics compared with 30–35 per cent of those from Northern Ireland. For Chemistry the corresponding figures were roughly 30 per cent and 20 per cent respectively. About 12 per cent of the pupils were following combined science courses of some kind, fewer than 10 per cent (mostly girls) were studying Human Biology, three to five per cent (mostly boys) were studying Rural Science and fewer than two per cent SCISP, Physics with Chemistry, Technology or Electronics (fewer than one per cent in this case).

There was clear evidence of the common practice in schools of 'channelling' pupils of different abilities into different science courses. The main sciences were more often being studied by pupils of average and high ability than by the least able pupils, while General Science, Human Biology and Rural Science were predominantly studied by average and less able pupils. Indeed, roughly a third of the most able pupils study all three main sciences, and the majority of these pupils study all three subjects to O-level. In contrast, the average and less able pupils usually study one, or at most two, science subjects.

The familiar tendency for girls in greater numbers than boys to choose Biology or Human Biology and for boys in greater numbers than girls to choose the physical and technological sciences persists. But there are indications that this traditional polarization is weakening, mainly because of the increasing take-up of Physics and Chemistry among girls.

Finally, it is interesting to note that the overall differences between the three countries in the proportions of boys and girls studying science can be attributed to take-up differences among average and less able pupils. Take-up rates for the main sciences were similar among the most able pupils in England, Wales and Northern Ireland among boys *and* girls. This finding is worthy of comment since the school system in Northern Ireland is still predominantly selective, and contains high proportions of mixed *and* single sex schools. The school systems in England and Wales are in contrast almost totally comprehensive and coeducational. The evidence is that the polarization between boys and girls into physical and biological sciences, respectively, is not exacerbated among able pupils by coeducational comprehensive schooling, as feared by some. The move to the comprehensive system in England and Wales would rather seem to have benefited average and less able pupils, by providing these pupils with new access to science subjects and courses previously unavailable to them.

3

Pupils' scientific interests and attitudes

3.1 Introduction

The previous chapter has served to illustrate once again the now familiar Biology/Physics polarization in the option choices of boys and girls, though this is weakening slightly year by year. But why does this arise? Why do girls in greater numbers than boys still study Biology in preference to Physics (or the technological sciences) when these subjects become optional? Indeed, why do boys in such large numbers 'drop' Biology and choose to embark on examination courses in Physics and Chemistry – science subjects generally considered by pupils as 'difficult' at this level? Many educators and researchers have addressed questions such as these in recent years, sharing a particular concern about the ways in which the pupils' option choices narrow their science education, and in the case of girls tend to limit later career opportunities. A number of special reports have focused on this issue: see, for instance, DES, 1980; Royal Society and Institute of Physics, 1982; Kelly ed, 1982; Hannan *et al*, 1983; Kahle ed, 1985; and Johnson and Murphy, 1986.

Different hypotheses have been proposed to explain both the subject polarization, and girls' relative weakness compared with boys in Physics tests. Essentially these possible explanations revolve around the nurture/nature argument. It has been suggested that girls are innately less able than boys in the kinds of skills and levels of conceptual understanding involved in Physics (and Mathematics) learning. Hence their early-established relative weakness in Physics and some aspects of Mathematics, and their later aversion towards these subjects and the occupations in which they are important.

The competing hypothesis focuses on socialization influences. The argument is that boys and girls from a very early age perceive and begin to emulate the traditional occupational pattern among men and women. These perceptions shape their role expectations as adults and have an immediate influence in delineating the kinds of activities and interests they consider most appropriate to them as members of a particular sex. Thus, not only do they begin to form sex-stereotyped views about particular jobs, but they also make similar judgements about voluntary hobbies and activities and, later, about the value and appropriateness to themselves of difference science and other school subjects.

In an attempt to throw further light on this issue, the 15 year olds involved in the 1984 survey were asked to respond to an extensive questionnaire about their *subject interests* and *job perceptions* (also an enquiry in the corresponding survey at age 13). The pupils were given the questionnaire after they had finished the particular test package which had been assigned to them. In fact, there were two questionnaires, each given to half the total sample. Between them they embodied seven different enquiries relating to the pupils' views about their science subjects, to their perceptions about the sex appropriateness of particular jobs, to their opinions about the suitability of different jobs for themselves, and to their awareness of, and interest in, topical scientific applications.

3.2 Pupils' attitudes towards their science-option choices

All the pupils who took part in the 1984 survey were asked for their views about their school subjects. Specifically, they were presented with a fairly comprehensive list of 21 named subjects grouped into languages, sciences and 'miscellaneous' (each with the option to add additional subjects not listed by name – see Appendix 2 for full details). The pupils were asked to tick grid boxes appropriately to indicate which subjects they were currently studying, and to give some reason(s) for their choices. In addition, they were asked to indicate whether or not they were enjoying studying the subjects concerned, and again to tick one or more possible reasons for their views.

It is interesting to note that of all the possible permutations of the 21 named school subjects in the list, no single combination was being studied by as many as 1 per cent of the pupils in the entire survey sample. The *most* popular combination was the following: English, French, Biology, Chemistry, Physics, Geography, History and Mathematics. These eight subjects were being studied by about 0.6 per cent of the 13,000+ pupils. This particular chapter will focus on the information available about the pupils' *science* choices, with only occasional comments about other subjects.

In answer to the question 'why did you choose to study this subject?' the pupils were given four possibilities:

'it's interesting', 'useful for jobs', 'it's easy' and 'other reason'. They were invited to tick one or more of these options in response. The results of this particular enquiry are revealing. To provide a point of reference for comparison, the percentages of pupils giving each reason are shown in Table 3.1 for English and Mathematics in addition to the listed science subjects. English and Mathematics are, of course, compulsory subjects for all but a few of pupils at this age, so that in these cases the pupils' views would not in practice act as choice criteria.

It should be noted at this point that the various percentage figures given in Table 3.1 (and in Table 3.2) are based on very different, sometimes rather small, sample sizes. This is because the responses relate to those pupils currently following courses in the different subjects, and, as the previous chapter has shown, take-up rates vary enormously from one subject to another. For example, the three main sciences are each studied by a third or more of all 15 year olds, but this is not the case with other subjects. In particular, as the previous chapter has shown, fewer than one per cent of pupils at this age study Technology or Electronics. There are, moreover, some striking differences in the proportions of boys and girls studying the various subjects, and in the proportions of pupils of different general academic ability doing so.

This is relevant to Table 3.1, which shows that Biology and Human Biology were the science subjects chosen most frequently because the pupils concerned found them interesting. It should be noted, though, that for *every* science subject more than half the pupils studying the subject considered it interesting at the time of option choices. Among the science subjects Physics was the subject most often agreed to have job value, followed closely by Chemistry. General Science, Human Biology and Rural Science were at the bottom of the list on this criterion. According to the evidence in Table 3.1 few pupils considered any science subject to be 'easy'.

These results are very much in line with those reported by Ryrie *et al*, 1979, on the basis of a smaller scale enquiry among pupils in Scotland. Interest in a subject and perceived usefulness for jobs were both the most common reasons given by pupils for their subject choices. The evidence is that Physics and Chemistry were chosen for their occupational value more frequently than other science subjects. Finding a subject 'easy' was a reason which featured very much less prominently among the pupils' responses.

There were some differences between boys and girls in terms of their subject perceptions, as Table 3.1 shows. Proportionally more girls than boys considered Human Biology and Rural Science to be interesting at the time they made their option choices. In contrast, proportionally more boys than girls claimed to have chosen to study Electronics and Technology for this reason (though extremely small numbers of girls here). There were only small differences between boys and girls for other science subjects on this criterion, including, interestingly, Physics.

The greatest differences between the percentages of boys and girls claiming to consider a subject 'useful for jobs' emerged for Electronics and Technology, followed by Physics and Rural Science. In each case greater proportions of boys than girls claimed to have chosen the subject for its occupational value. The science subjects chosen for this reason by slightly higher proportions of girls than boys were Human Biology, Biology and Chemistry. However, the differences here were slight.

As far as 'easiness' of subjects is concerned, it is interesting to note the tendency throughout the list of science subjects for slightly higher proportions of boys than girls to offer this as a reason for subject choice. This tendency persists in their current views about their chosen science subjects and in their views about English and Mathematics. This evidence of a greater level of confidence on the part of boys in their own abilities has been reported previously for Mathematics at this and younger ages, and has been shown to be unjustified on the basis of their comparative test performances (see, for instance, the report of the 1979 APU Mathematics Survey – Foxman *et al*, 1981a, or the APU Mathematics Review Report – Foxman *et al*, 1985).

Table 3.1 *Pupils' reasons for their option choices in science*
(Percentage of pupils giving each reason among those studying the subject – more than one reason possible)

	Number of pupils		'Interesting'		'Useful for jobs'		'Easy'	
Subject	Boys	Girls	Boys	Girls	Boys	Girls	Boys	Girls
English	6,291	6,453	24	32	64	67	7	6
Mathematics	5,919	6,113	27	22	66	66	7	4
Physics	3,551	1,433	48	46	63	52	4	2
Chemistry	2,224	1,767	53	49	51	53	5	1
Biology	1,329	4,617	73	72	32	36	8	4
General Science	958	720	59	57	27	26	10	7
Human Biology	530	970	62	71	23	29	9	4
Rural Science	452	167	55	63	25	14	14	9
Electronics	297	58	56	22	41	10	7	2
Technology	649	68	56	43	45	19	9	3

Table 3.2 *Pupils' current views about their subjects**
(Percentage of pupils claiming to hold each opinion among those studying particular subjects)

Subject	'Enjoyable'		'Interesting'		'Difficult'		'Useful for jobs'	
	Boys	Girls	Boys	Girls	Boys	Girls	Boys	Girls
English	70	83	34	46	6	5	54	58
Mathematics	59	55	30	24	13	24	56	53
Physics	64	55	47	44	22	36	48	36
Chemistry	61	58	49	46	24	35	38	36
Biology	77	78	67	68	11	16	30	32
General Science	66	64	50	50	7	11	26	23
Human Biology	62	75	55	60	12	12	25	30
Rural Science	67	60	47	52	6	8	24	14
Electronics	63	28	50	33	12	19	40	12
Technology	70	49	50	37	8	13	42	19

*For sample sizes see Table 3.1.

Unfortunately it has not been possible to investigate the phenomenon directly with science data.

When asked whether they were enjoying studying their 'chosen' science subjects, between 60 per cent and 80 per cent agreed that they were (a lower 57 per cent for Mathematics). In every case the proportions claiming to enjoy their studies were higher than those claiming originally to have chosen them because they were interesting. On the other hand, with the exception of English, the proportions *still* considering their subjects to be interesting were slightly lower than before, as Table 3.2 shows.

Greater proportions of pupils currently considered their science subjects to be interesting than felt the same about English or, particularly, Mathematics. On the other hand, over half the pupils still considered English and Mathematics useful for jobs, compared with just under half for Physics and a third to a quarter for the other science subjects. On the evidence of Table 3.2, Chemistry and Physics were more frequently perceived by the pupils to be 'difficult'. This is an expected finding in the light of previous research – see Ormerod with Duckworth, 1975. More than a quarter of the pupils studying these two subjects considered them to be 'difficult' compared with 19 per cent for Mathematics, 14 per cent or fewer in the case of other science subjects and just five per cent for English.

It is not possible on the basis of this survey data to say whether this perceived difficulty is a consequence of the examination courses the pupils follow at this age or whether this view already prevailed. It might be that many of these same pupils already felt this way before they made their option choices, and that they chose these subjects in spite of their 'difficulty' – perhaps for their high job value.

Once again there are differences in the response patterns for boys and girls, as Table 3.2 shows. However, only for English, Electronics and Technology were there large differences in the percentages of boys and girls claiming interest: in favour of girls for English and in favour of the boys for Electronics and Technology. On the criterion 'useful for jobs' the largest opinion gaps are all still in favour of the boys: for Electronics, Technology, Physics and Rural Science. Only slightly higher proportions of girls than boys considered English, Biology and Human Biology to be of value in future occupations.

For every science subject except Human Biology lower proportions of boys than girls claimed to find the subject difficult. Of especial interest are the particularly large differences in the proportions of boys and girls who thought Physics and Chemistry to be difficult: roughly a quarter of the boys studying these subjects compared with over a third of the girls. Mathematics engendered a similar response difference; a quarter of the girls claimed to find the subject difficult – twice the proportion of boys doing so.

Evidence is available in the case of Mathematics which shows this subject already to be considered 'difficult' by proportionally more girls than boys at age 11, this perception strengthening for both sexes – but more for girls than boys – through the secondary school (see Foxman *et al*, 1981a, 1982). Moreover, the topics within Mathematics for which the greatest differences of opinion arose at the earlier age were just those most relevant to Physics – in particular 'Measures' (see Foxman *et al*, 1981a, 1981b). It is possible that the large differences in the proportions of 15 year old boys and girls who consider Physics (and Chemistry) to be difficult would already be present among pupils at age 13 or younger.

To summarise some salient findings, it seems that much higher proportions of the boys compared with the girls who were studying Electronics and Technology found these subjects interesting and considered them to be useful for jobs. Also, proportionally fewer of the boys thought these subjects 'difficult'. For Mathematics proportionally more boys than girls claimed to find the subject interesting, and proportionally fewer thought it difficult.

For Chemistry and Physics the picture is a little different, in that similar proportions of the boys and girls thought these subjects interesting despite the fact that markedly higher proportions of girls than boys found them difficult. While Chemistry was perceived by similar proportions of the boys and girls studying this subject to have some 'job value' for themselves, this was not the case for Physics which, along with Electronics and Technology, was considered 'useful for jobs' by much larger proportions of boys than girls. If similar differences in the proportions of boys and girls holding these views *do* also prevail among 13 year olds – as is likely – then little wonder fewer girls than boys accept the challenge of a Physics examination course by allowing personal 'job value' to override 'subject difficulty'. Unfortunately the pupils in *this* survey were not asked about their perceptions of subject 'difficulty' at the time they made their option choices.

3.3 Pupils' job perceptions

Why *should* Physics, Electronics and Technology be considered by greater proportions of boys than girls to have some 'job value' for themselves? Is it that these subjects are more relevant to the specific *kinds* of jobs boys are more likely than girls to have on leaving school/college? Or is their usefulness perceived to be more general than this, applying to a wide range of occupational possibilities?

It is well known that boys and girls tend to follow the established pattern of employment in their own job aspirations (see, for instance, Ryrie *et al*, 1979, Pratt *et al*, 1984). Many girls tend to look to servicing and caring occupations, which have traditionally been dominated by a female workforce. Examples are catering, secretarial work, nursing, and so on. Many boys, on the other hand, set their sights rather on those occupations which have traditionally been the work of men. These include jobs in the skilled and semi-skilled trades (plumbing, joinery, electrical, and the like), engineering and building. In most occupations, including the professions, men more often than women are found in managerial and supervisory roles. The pupils' personal aspirations *within* these broad bands will, of course, depend very much both on their own predilections and on their real and perceived abilities. And indeed it is likely that among the more able pupils many of both sexes will share similar career intentions.

The set of enquiries about pupils' job perceptions included in the survey questionnaires was based on a list of 44 named occupations (for details see Appendix 2). These included those jobs *expected* to be seen as uniquely 'male appropriate' or as 'female appropriate' by the pupils (bricklayer, plumber, secretary, nurse, and so on), as well as others which it was thought might be considered more neutral in this respect (teacher, architect, computer programmer, etc). In addition, where possible, jobs were included which represent different hierarchical levels, and which therefore carry differential status, within a particular occupational field (an example is bank manager and bank clerk). The expectation here was that the higher status jobs would be considered more 'male appropriate' than the lower status jobs – again, attitudes which would reflect the prevailing employment pattern.

The question about 'sex appropriateness' asked the pupils to indicate the suitability of each job first for women and then for men, using a five point scale ranging from 'very suitable' to 'not at all suitable' (the pupils simply checked the appropriate grid box). Some jobs – such as 'teacher', 'university lecturer' and 'shop assistant', would be difficult to judge in this way because of their inbuilt ambiguity. Since the pupils were not given the opportunity to qualify their responses, for these and a number of other jobs the task asked of them was not a straightforward one. Nonetheless, the general outcome is clear enough.

The results of this enquiry are very much as might be expected from previous knowledge of pupils' job perceptions. Moreover, boys and girls were in general accord in their views. With the exception of 'teacher' and 'journalist', *every* job in the list was perceived by the pupils to be more appropriate for men or for women. There is no evidence in the data that pupils of one sex have more stereotyped views than those of the other. There *was,* though, a definite tendency for girls to consider *every* job to be more suitable for both men *and* women than boys.

Rather than present full data for the entire job list here (this is given in Appendix 3), it will suffice to list the most severely sex stereotyped jobs, along with the most neutral in this respect. As Table 3.3 shows, few pupils of either sex were willing to consider hairdressing, nursing or secretarial work to be suitable for men, or engineering, bricklaying, farming, plumbing, garage jobs, electrical work or – on the sole evidence of 'steelworker' – heavy industry to be appropriate occupations for women.

Table 3.3 *Pupils' ideas about the 'sex appropriateness' of different jobs*

(Percentage of pupils considering job 'very suitable' for women or men – roughly 5,000 pupils in total)

For women	For men	Occupation
1–10	65–70	Engineer, Bricklayer, Garage Mechanic, Farmer, Steelworker, Plumber, Electrician
41–42	46–47	Teacher, Journalist
75–85	10–20	Hairdresser, Nurse, Secretary, Typist

The imbalance in the numbers of jobs in each group in Table 3.3 is typical of the entire list. Of the 44 named occupations just 13 were considered to some extent to be more suitable for women than men, and 29 were deemed to be more suitable for men than women. In the former group, in addition to the most severe examples shown in the table, we find – in order of increasing 'equivalence' of suitability – librarian, shop assistant, office cleaner, café worker, cashier, social worker, clerical worker, post office worker and shopkeeper. Among the additional jobs considered more appropriate for men than women we have farm worker, bank manager, Member of the Forces, driver, factory manager, professor, police officer, porter, petrol pump attendant, machinist, caretaker, architect, doctor, scientist, solicitor, factory worker, computer programmer, Member of Parliament, bank clerk, university lecturer, chef and lab technician.

There were *seven* instances where jobs were included at different 'status' levels in the same occupational field: bank manager and bank clerk, factory manager and factory worker, professor and university lecturer, scientist and lab technician, farmer and farm worker, shopkeeper and shop assistant, doctor and nurse. For every pair, the perceived suitability for men rather than women increased with increased job status.

There is clear evidence in the responses of pupils of different abilities that sex stereotyping increased with decreasing ability for those jobs already identified as most strongly sex stereotyped. Thus, proportionally more of the least able compared with the most able pupils considered the following occupations to be 'not at all suitable' for women: plumber (50 per cent and 33 per cent, respectively), farmer (35 per cent, 19 per cent), engineer (38 per cent, 15 per cent), garage mechanic (49 per cent, 37 per cent) and electrician (38 per cent, 16 per cent). For those jobs generally considered to be the most 'female appropriate' this view was again more prevalent among the least able pupils. Proportionally more of the least able compared with the most able pupils thought the following jobs 'not at all suitable' for men: secretary (26 per cent compared with 19 per cent), typist (34 per cent, 21 per cent), nurse (27 per cent, 11 per cent) and hairdresser (16 per cent, 8 per cent).

Turning attention to the pupils' opinions of the suitability of the various jobs for *themselves*, not surprisingly we find a strong association with their views about 'sex appropriateness'. This is combined with a strong distinction between those jobs which the most able and the average or less able pupils, respectively, might realistically aim for. Thus, almost half the least able boys considered 'driver' a very suitable job for themselves, about a third felt the same way about 'bricklayer' and 'garage mechanic', and around a quarter shared this view with regard to 'farmer', 'engineer', 'electrician' and 'Member of the Forces' (compared with 10 per cent or fewer of the most able boys in each case, with the exception of 'engineer' at 18 per cent – another indication of the dual image of engineering). Among the most able boys 'scientist' and 'computer programmer' joined 'engineer' in attracting the highest percentages of 'very suitable for self' responses (at just under 20 per cent).

Among the least able girls, about a third considered 'hairdresser', 'nurse' and 'shopkeeper' to be very suitable occupations for themselves in principle, while a fifth to a quarter felt the same way about 'secretary' and 'typist' (these two were unusual in attracting the greatest percentages of 'very suitable' responses from average ability girls). The highest percentage of 'very suitable' responses among the *most able* girls emerged for 'journalist' at 18 per cent. This was followed by 'teacher', 'nurse' and 'social worker' at 13 per cent; all other occupations attracted 10 per cent or fewer 'very suitable' responses. With the exception of 'teacher' and 'journalist', among the most able pupils fewer girls than boys considered any of the listed professions as 'very suitable' for themselves.

In the enquiry into their ideas about the relevance of Physics and of Biology to different occupations, the pupils were asked to rate the importance to these subjects to each job according to the usual five point scale – this time the scale ran from 'very important' to 'totally unimportant' (see Appendix 2). As mentioned earlier, it is important to bear in mind that in all these job perception enquiries the pupils were not given the opportunity to qualify their responses in any way. For example, they were not in any position to distinguish, say, a shop assistant in a DIY store from one in a sweet shop when considering the value of Physics to the work. In this sense, for some of the listed occupations, the task asked of the pupils here was again rather an artificial one. Moreover, it should be recognised that the pupils' responses will not always have been based in informed opinion. Nonetheless, this enquiry has also produced interesting findings – some expected, others less so.

The detailed results for all 44 different listed jobs are given in Appendix 3. Once again the boys and the girls were in general agreement about the degree of usefulness of the two sciences to these jobs, and many of the jobs can be grouped together in respect of the pupils' assumptions. In particular, the largest such group contains all those jobs for which the majority of boys *and* girls considered neither Biology nor Physics to be particularly necessary. These are: bank manager and bank clerk, shopkeeper and shop assistant, factory manager and factory worker, MP, solicitor, journalist, librarian, secretary, clerical worker, typist, cashier, post office worker, social worker, driver, petrol pump attendant, porter, caretaker, café worker, bricklayer and office cleaner.

Fewer than one in twenty pupils considered Biology or Physics to be 'very important' for these jobs; indeed, typically half or more of the pupils considered these subjects to be 'totally unimportant' in these cases. Where slight differences emerged in the responses of boys and girls these reflected a tendency for girls to be marginally more persuaded than boys of the general value of both science subjects, but particularly Biology, for most jobs. Typically there would be smaller proportions of girls considering the subjects to be of no importance at all for the jobs concerned. Only in the case of 'hairdresser' were the differences of opinion large enough to be especially noteworthy. While half the boys considered neither science subject to be of any importance at all in this job, just a quarter of the girls shared this view for Biology and a third for Physics (it must be remembered that few pupils of either sex considered either subject to be 'very important' or even 'quite important' for this or the other jobs mentioned earlier).

The discriminating factor among the remaining jobs tended to be the *assumed* importance of Physics, as Table 3.4 shows. Throughout the list there was a definite tendency for Physics to be considered by both boys *and* girls to be more relevant than Biology, a realistic perception in the light of employers' and educators' views for these particular jobs. Indeed, only for 'scientist', 'lab technician', 'doctor' and 'nurse' do most pupils consider Biology to be 'very important'. Biology and Physics were considered roughly equal in importance for scientists and laboratory technicians and also – though in these cases of much less general importance – for university lecturers and professors!

Table 3.4 *The importance attached by 15 year olds to Biology and Physics for particular jobs*
(Percentage of pupils indicating extreme views – roughly 5,000 pupils in total)

	'very important'		'not at all important	
Occupation	Biology	Physics	Biology	Physics
Doctor	86	47	3	7
Nurse	72	29	7	16
Farmer	32	12	12	24
Scientist	80	83	4	5
Lab technician	52	60	10	7
Professor	46	50	13	10
University lecturer	27	31	17	15
Electrician	3	46	47	14
Engineer	3	40	44	16
Computer programmer	2	24	47	16
Architect	5	20	40	23
Garage mechanic	3	25	56	21
Plumber	1	13	53	23
Machinist	1	17	54	24
Member of Forces	5	9	30	26
Steelworker	2	14	55	27

As noted earlier, the girls were in general more inclined than boys to the view that both science subjects – but particularly Biology – have *some* usefulness in all the occupations. The largest difference of opinion between them concerned the value of Biology to nurses (63 per cent and 80 per cent, respectively, of boys and girls considered Biology to be 'very important' in this job). Indeed, nursing is the only 'female' occupation listed in Table 3.4; it is the only female occupation for which at least half the pupils considered either science subject to be of *some* importance.

The assumed importance of Physics to those traditional 'male appropriate' jobs, especially 'engineer', 'electrician' and 'mechanic', is noteworthy. Only in the case of 'farmer' and 'doctor' do we see examples of 'male' occupations for which Biology is perceived to have more relevance than Physics. Also interesting in Table 3.4 is the greater importance attached to both sciences for the higher status, more 'male appropriate', jobs in related pairs.

Here then is evidence that both boys *and* girls consider Physics to be more useful than Biology in a wide range of those possible occupations they both view as most appropriate for men rather than women. Little wonder that proportionally fewer of the girls compared with the boys who were currently studying Physics claimed to have chosen this subject because of its future job value.

3.4 Opinions about various scientific applications

It is of some interest to know to what extent pupils at this age are aware of the contribution applied science makes to society. In particular it would be interesting to hear their views about the benefits or drawbacks of some of the more widely publicised applications, one or two of which are seen as controversial issues among the public at large. In order to sample their views, the pupils were presented with a list of scientific applications, some medical, some biological, some military, and most considered currently newsworthy by the media. The pupils were asked whether they had heard of the application at all, if so whether they found it interesting and whether they considered it to be a 'good' thing or a 'bad' thing for society – with reasons. 'Can't say' was an optional response in every case.

For ease of communication of the results, the applications have been subdivided into four groups, as shown in Table 3.5 (p22). The first group contains those medical applications of which the vast majority of these 15 year olds, irrespective of sex, claimed some awareness – heart transplants, test tube babies and cancer research. The table shows clearly the greater appeal of these particular applications to girls than to boys. The second group contains the only other two applications (related as it happens) for which pupil awareness was similarly extremely high for both sexes – nuclear power and nuclear weapons. Boys and girls shared similar interest levels in these.

The third group comprises the four technological applications with which fewer – but still high proportions of – pupils claimed some familiarity. For each of *these* applications 'awareness' was higher among the boys than the girls, and more boys than girls claimed interest. The fourth and last group contains those applications least familiar to the pupils; with the exception of factory farming, it seems that the girls were again less well informed and they again claimed less interest than the boys.

The rather striking polarity in the interests of the boys and girls in these various issues follows expectations to a great extent. After all, it has been found previously that, from an early age, girls tend to be most interested in biological topics and 'health' while boys are more interested in space science and physical science/technology (see, for instance, Harvey and Edwards, 1980, Lie and Bryhni, 1983, Ormerod and Wood, 1983, Smail and Kelly, 1984).

The differences in awareness levels between boys and girls might be of some concern. With such newsworthy, even in some cases controversial, applications it is difficult to accept that girls can, as a group, be less exposed to the issues than are boys. On the other hand, boys have at all ages surveyed claimed to be more avid television watchers and book/magazine readers than girls where science is concerned. In an enquiry in the corresponding survey of 11 year olds, for instance, almost twice the proportion of boys as girls claimed 'quite often' to watch science and science fiction programmes on television (roughly half and a quarter of the pupils, respectively, in the case of science fiction).

Discrepancies in favour of boys also emerged for reading science and science fiction books, although the differences here were smaller (see Russell *et al*, 1988 or Johnson and Murphy, 1986, for details). Presumably this greater media exposure to science fact and fiction over the years has a cumulative effect in broadening boys' knowledge base and, perhaps, in widening their sphere of interest. The situation revealed in Table 3.5 is one consequnce of this.

Table 3.5: *Awareness of, and interest in, various scientific applications**

(Percentage of pupils responding in indicated ways)

	'Heard of'*		No. pupils		'Interested in'	
Application	Boys	Girls	Boys	Girls	Boys	Girls
Heart transplants	96	97	2,880	2,910	52	80
Test tube babies	95	97	2,850	2,910	38	78
Cancer research	93	96	2,790	2,880	63	87
Nuclear power	94	93	2,820	2,790	55	47
Nuclear weapons	94	94	2,820	2,820	60	53
Space exploration	89	85	2,670	2,550	66	41
Cable television	92	82	2,760	2,460	69	52
Satellite communication	86	76	2,580	2,280	62	38
Robotics	85	79	2,550	2,370	59	38
Laser technology	76	62	2,280	1,860	64	39
Factory farming	62	63	1,860	1,890	31	27
Intensive agriculture	46	38	1,380	1,140	34	27
Genetic engineering	48	40	1,440	1,200	51	32
Fibre optics	43	18	1,290	540	55	34

* 5,900 pupils in total – roughly half boys, half girls; the percentages in the 'interest' columns are based on the numbers of pupils who had heard of the application concerned. These are shown alongside.

As to opinions about the value of each application to society, the boys and girls were in *general* agreement. They shared positive attitudes towards the potential value of most of the applications, as Table 3.6 shows. Only for nuclear weapons (and, incidentally, 'acid rain'– not an application but rather a side effect) was their view almost unanimously a negative one, with 70 per cent of the boys and the girls considering these to be 'bad' for society. The reasons given by the pupils for their views might be anticipated: nuclear weapons are 'bad' because they 'endanger innocent people', 'could start a war', 'will destroy the world' and so on; they are 'good' because they 'keep peace', 'countries would be defenceless without them' and 'will be a defence in World War 3'!

Table 3.6: *Opinions about the value to society of each scientific application*

(Percentage of pupils giving each response among those claiming awareness)

	No. pupils		'Good thing'		'Bad thing'	
Application	Boys	Girls	Boys	Girls	Boys	Girls
Heart transplants	2,880	2,910	87	92	3	2
Test tube babies	2,850	2,910	57	74	17	9
Cancer research	2,790	2,880	89	91	2	1
Nuclear power	2,820	2,790	36	17	39	51
Nuclear weapons	2,820	2,820	15	9	69	71
Space exploration	2,670	2,550	68	56	6	6
Cable television	2,760	2,460	64	49	12	15
Satellite communication	2,580	2,280	81	60	2	3
Robotics	2,550	2,370	43	24	22	21
Laser technology	2,280	1,860	62	47	6	4
Factory farming	1,860	1,890	33	23	30	30
Intensive agriculture	1,380	1,140	51	43	15	7
Genetic engineering	1,440	1,200	36	22	21	14
Fibre optics	1,290	540	66	40	3	3

In the case of nuclear power, on the other hand, the pupils were about evenly split between those holding positive views, those holding negative views and those who were undecided. Again, among those with definite views the reasons given for holding these could be anticipated. Many of the pupils who thought nuclear power a 'bad' think usually gave reasons such as 'kills people', 'dangerous', 'encourages war' (this one particularly suggestive of an inability to distinguish nuclear power from nuclear weapons). There were some pupils who explicitly mentioned the possibility of radioactive leaks – and this before Chernobyl.

Factory farming attracted negative responses from 30 per cent of the pupils, and robotics attracted a negative response from about one in five pupils. In the case of robotics the overwhelmingly most popular reason given for the view that this application is a 'bad' thing was the familiar one 'puts people out of work'.

Among the three well known medical applications only 'test tube babies' engendered doubt about its worth in some pupils' minds – interestingly, more of these pupils were boys than girls. The pupils' reasons for their negative views revolved around the 'interfering with nature' argument. Many were inspired by a fear of 'something going wrong' and of 'monsters' being created. A handful of pupils objected on moral grounds, showing concern for the 'discarded embryos'.

The three medical applications are the only ones in the entire list which more girls than boys considered to be 'good' things. For all other applications the boys emerged the more favourably disposed, in most cases strikingly so. A glance through Table 3.6 will confirm this, with differences typically of 10 to 20 percentage points in the proportions of boys and girls considering the application to be 'good'.

Among pupils of different abilities there were some interesting and general differences. The more able the pupils the more likely they were to have heard about the application, the more interested they claimed to be in it and the more likely they were to have a definite view one way or the other on its value to society. Typically, 85–100 per cent of the most able pupils had heard about most of the applications in the list. The only exceptions to this were factory farming and intensive agriculture (68 per cent and 66 per cent respectively), genetic engineering (59 per cent) and fibre optics (46 per cent).

The applications most well known among the *least able* pupils were 'heart transplants', 'test tube babies', 'cancer research', 'nuclear power' and 'nuclear weapons' (all between 80 per cent and 90 per cent). The least well known applications among this group were the four already indicated above for the most able plus 'laser technology'.

Only for 'factory farming' and 'intensive agriculture' were the proportions of pupils expressing interest higher among the least able than among the most able pupils (though differences of just four or five percentage points were involved). These two applications show another unique feature in terms of the pupils' views about value to society. Among the least able group similar percentages considered both applications to be 'good' (38 per cent and 43 per cent respectively) and similar percentages considered both to be 'bad' (15 per cent and 11 per cent respectively). Among the *most able* pupils, on the other hand, the percentages considering factory farming and intensive agriculture to be 'good' things were 21 per cent and 50 per cent, respectively; and the percentages thinking these to be 'bad' things were 41 per cent and 12 per cent, respectively. Thus, it seems that the more able the pupils the more likely they were to discriminate between factory farming and intensive agriculture, and to perceive these as different kinds of application, the one involving animals – with consequent concerns about animal well being – the other involving plants.

3.5 Summary

The reasons most often indicated for their science option choices by the 15 year olds questioned in this survey were that the subjects concerned were interesting or useful for jobs. Biology was the subject chosen by the greatest proportion of pupils for its interest value. Physics and Chemistry were preceived to be the science subjects with the highest 'job value', following behind English and Maths.

Among boys and girls there were some strong differences of opinion. Of those few pupils who were studying Electronics and Technology, proportionally more of the boys than the girls chose these subjects for their interest value. Physics, Electronics and Technology were considered 'useful for jobs' by proportionally more boys than girls. Proportionally more girls than boys claimed to find Physics, Chemistry and Maths difficult.

Boys and girls were in general agreement about the 'sex appropriateness' of particular occupations and about the value of Biology and of Physics to these. Teaching and journalism are alone among a list of 44 different occupations in being considered by boys and girls to be equally suitable for men and women. The most sex-stereotyped 'male' jobs would appear to be those of engineer, bricklayer, plumber, garage mechanic, electrician, steelworker and farmer. The most sex-stereotyped 'female' jobs are those of nurse, hairdresser, secretary and typist. All these jobs were considered 'very suitable' for themselves by much greater proportions of average and less able pupils (typically a quarter to a third) than by the most able pupils (typically fewer than 10 per cent) of the appropriate sex. The most able pupils favoured the 'professions', with proportionally more boys than girls considering every professional occupation except 'teacher' and 'journalist' to be 'very suitable' for themselves.

The only 'female' occupation for which at least half the pupils considered Biology or Physics to be of some value was nursing. In contrast, Physics was considered by the majority of pupils to be of some importance to all the most 'male appropriate' occupations. The only such occupations for which Biology was considered of equal or greater importance compared with Physics were 'farmer' and 'doctor'.

The pupils' degree of awareness of, and interest in, topical scientific and technological applications varied with ability and sex. The more able the pupils the more likely they were to have heard about a range of named applications, the more interested they claimed to be in each and the more likely they were to hold a view – positive or negative – on their value to society. On the whole, the evidence is that boys are better informed about a wider range of applications than are girls, and their interests differ appreciably also. Girls were, as a group, more interested than boys in the medical applications (cancer research, test tube babies, heart transplants) and boys were more interested than girls in the physical science applications (such as laser technology, fibre optics, cable television, space exploration).

4

Pupils' levels of science performance

4.1 Introduction

A relatively small and shrinking proportion of pupils 'drop' all science at 13+, as Chapter 2 has shown. The great majority of 15 year olds study at least one science subject. On the other hand, while the three main sciences continue to predominate in the science curriculum at this age, the variety in the science experience of pupils is considerable. Individual pupils might be studying any combination of the major science subjects, from a single main subject or general science in the case of the majority of average ability pupils, to all three main sciences in the case of the most able. As Chapter 3 has shown, the pupils studying the various subjects generally enjoy studying them, and most are considered to have some job-value by many pupils. Physics and Chemistry – along with Mathematics – are considered to be relatively 'difficult'. Given this background, what do we know about the pupils' *performances* in science?

As indicated in Chapter 1 a number of different aspects of science activity have been identified as important to assess within the context of this monitoring exercise. These aspects are articulated as the categories and subcategories of the assessment framework. The kinds of questions developed to represent each subcategory and the mode of assessment adopted in each case are fully described in previous survey reports at this and the other two ages (11 and 13) and also in the chapters which follow.

Domain-sampling is a term used to describe the strategy by which questions are randomly selected from within a question pool to represent a particular subcategory in a survey. Pupil performance on the collection of selected questions is assumed to reflect their likely performance on the whole 'domain' of questions of similar type. Thus the sample of questions drawn in any year from, say, the pool for **Interpreting presented information** is assumed to represent the 'population' of questions which could similarly be classified into this subcategory, in the same way that the sample of 15 year olds attempting these questions is assumed to represent all the pupils of this age in ordinary secondary schools. The difference in practice is that the population of 15 year olds is 'real', ie these pupils exist and can be identified, whereas the population of questions does not exist but can only be imagined.

In some cases, for economic and logistic reasons, it has not proved possible to adopt a domain-sampling strategy. This applies particularly to two of the three *practical* categories **Use of apparatus and measuring instruments** and **Performing investigations**. In the former case early attempts to establish a large pool of questions suggested that this was an artificial exericse, and that relatively few questions sufficed to cover the range of typical items of apparatus and measuring instruments which 15 year olds would be expected to be able to use. It also proved impossible readily to implement a random selection of questions, since the apparatus and supervisory load on the administrators would have been too great. This subcategory is therefore now represented by a limited number of questions, all of which are administered in any survey (see Chapter 6 for details). Pupil performance is reported in detail for each separate question.

Performing investigations (along with its *written* counterpart **Planning entire investigations**) is also represented by a small number of questions each time, with pupil performance reported in detail at the question level (see Chapter 11). In this case cost constraints are especially severe: each practical 'task' engages individual pupils for up to the full hour of available testing time, and testing is on a one-to-one basis with trained administrators carrying all necessary equipment to the schools. At most five pupils can be tested in one school visit, and just a handful of different questions have been administered in any particular survey.

For most subcategories in the assessment framework domain sampling of questions proved a possibility – at least in later years once the question pools had begun to approach an appropriate size. It is with *these* subcategories that this chapter is concerned.

The 1984 survey was the last in the initial series of five annual surveys, so that this is a particularly opportune time to review the cumulated findings for consistencies and trends. It is also an appropriate time to explore further some features which have emerged in earlier surveys, now that a greater amount of similar data is available for analysis. In this chapter we begin by presenting the performance results for the domain-sampled subcategories for the 1984 survey, before moving on to review comparative figures for the period 1980–84.

4.2 Pupils' subcategory performance levels in 1984

Seven subcategories at age 15 were represented in the 1984 survey by collections of questions randomly selected from relevant question pools (ie these subcategories were 'domain-sampled' to produce representative survey tests). These subcategories are (with the numbers of questions selected and the ways in which they were packaged shown in brackets):

- Using graphs, tables and charts (90–6×15)
- Making and interpreting observations (45–3×15)
- Interpreting presented information (66–6×11)
- Applying biology concepts (66–6×11)
- Applying chemistry concepts (66–6×11)
- Applying physics concepts (66–6×11)
- Planning parts of investigations (60–4×15)

The testing time usually demanded of any individual pupil at ages 13 and 15 is one hour. Early pilot studies indicated that around 15 questions is the most appropriate number to include in a test package occupying this time at age 13, whereas similar trials suggested that 15 year olds could cope with up to 20 or so questions in the period. The test packages presented to pupils in this survey vary between these limits.

The 45 questions representing **Making and interpreting observations** were randomly distributed among three practical test packages to be administered in 'circus' format (a 'circus' is a strategy by which a group of pupils is tested at the same time, pupils working individually on the test questions by circulating around tables or work benches on which are set up the various tasks and resources).

The 90 questions selected to represent **Using graphs, tables and charts** were randomly distributed among six different test packages of 15 questions each. The 60 questions from **Planning parts of investigations** were similarly randomly distributed among four test packages of 15 questions each.

Questions from the 'concept application' subcategories and from the sub-category **Interpreting presented information** were distributed among 11 different test packages in such a way that each such package contained two different subcategories, eg **Interpreting presented information** with **Applying chemistry concepts** or **Applying physics concepts** with **Applying biology concepts**. This mixing strategy was adopted in these cases to avoid the possibility that pupils who had not studied, say, Physics since they were aged 13 might be faced with an hour's worth of Physics questions.

About 12,500 pupils took a subcategory 'test' of one sort or another in this survey and an additional 3,000 pupils took part in linked research exercises. All pupils were drawn from around 500 schools; schools and pupils were selected by a random sampling procedure – see Appendix 4. All the written test packages were administered in every school, with just one or two pupils in any school taking a particular one of these. In total, approximately 440 pupils attempted each 'written' package. The practical circuses were administered in a random subsample of 135 schools in total, with two circuses in each school and a group of nine pupils in each circus. Each practical circus was eventually attempted by roughly 680 pupils drawn from about 75 schools (around 15 schools in each of Wales and Northern Ireland).

During processing of the pupils' responses, mean percentage 'test' scores are first calculated for each pupil. These are produced simply by averaging the series of individual question scores achieved by the pupil and expressing this average as a percentage of the maximum possible total mark. Estimates of the 'population' performance levels of each of a number of identifiable pupil groups (such as boys and girls) are computed on the basis of these individual scores, weighting appropriately to allow for imbalance in sample representation (see Appendix 4 and associated technical report for details). The performance level estimates for 1984 are given in Table 4.1. It should be noted that the standard errors (s.e.) quoted in this and the following table take account of pupil *and* school sampling errors, but not of measurement errors arising from question sampling – see Appendix 4.

A familiar feature in the survey data to date and which is clear again in Table 4.1 is the difference in the general levels of performance scores across the various subcategories. Specifically, pupils' performance levels are higher for **Using graphs, tables and charts** than they are for every other subcategory and their mean performance scores are lowest for the 'concept application' subcategories. This is not to say, of course, that symbolic communication is *necessarily* an inherently 'easier' or even better practised activity than all others in the

Table 4.1 *Pupils' subcategory performance levels in 1984**

(Mean percentage scores – weighted population estimates)

Subcategory		All	Boys	Girls	Eng	Wales	NI
Using graphs, tables and charts	mean	75.7	76.9	74.4	75.9	73.4	74.7
	s.e.	0.5	0.6	0.7	0.6	0.8	0.7
Making and interpreting observations	mean	49.8	48.5	50.8	50.1	47.9	46.7
	s.e.	0.5	0.5	0.5	0.5	0.9	0.7
Interpreting presented information	mean	43.9	45.4	42.4	44.1	42.6	41.5
	s.e.	0.6	0.8	0.7	0.7	0.8	0.8
Applying biology concepts	mean	32.4	32.4	32.6	32.5	31.9	31.0
	s.e.	0.5	0.6	0.6	0.5	0.7	0.6
Applying chemistry concepts	mean	31.8	32.7	31.0	32.0	30.5	30.1
	s.e.	0.5	0.7	0.6	0.6	0.8	0.8
Applying physics concepts	mean	31.3	35.2	27.1	31.3	31.6	29.3
	s.e.	0.5	0.7	0.5	0.6	0.7	0.7
Planning parts of investigations	mean	42.4	43.0	41.7	42.7	38.9	40.5
	s.e.	0.6	0.8	0.7	0.6	0.8	0.8

* Sample size details are given earlier, in the text.

assessment framework, nor that applying science concepts is inherently more difficult than most other activities. Pupils' are simply more or less successful in these activities as they are represented in the kinds of questions developed for this survey programme, and as they are awarded in the associated mark schemes.

As to the relative performances of boys and girls, Table 4.1 shows that the boys produced higher mean scores than the girls on every subcategory except **Making and interpreting observations** (for which the performance gap in favour of the girls reaches statistical significance) and **Applying biology concepts**. Indeed, the differences in favour of boys are large enough to reach statisticial significance in the cases of **Using graphs, tables and charts, Interpreting presented information** and **Applying physics concepts**. This is by now a familiar pattern; as we shall see later it replicates the findings of most of the earlier surveys in this series. It is not entirely unexpected either, in the light of the gender related polarization in option choices described in Chapter 2. More girls than boys do study Biology at this age, and three times as many boys as girls study Physics. These performance differences must therefore be attributable at least in part to curriculum influences. This possibility is explored in Chapter 13.

Turning attention to relative performance between countries, we see again that the pupils in Wales and Northern Ireland have generally failed to match the performance scores of their peers in England across the range of subcategories. The mean score for Wales for **Using graphs, tables and charts** is significantly behind that for England (in the statistical sense, and at the usual 5 per cent level), while the mean score for the Northern Ireland pupils when **Applying physics concepts** is significantly behind those of both England and Wales (while the former difference has been a consistent feature in the survey data, the latter has not). The Northern Ireland pupils have produced a significantly lower score than their peers in England on the practical 'observation' subcategory. Again, these national differences, small as they are, can plausibly be linked to, among other things, differences in the general levels of science take-up in the three countries (see Chapter 13).

As Table 4.2 shows, *within* England pupils in the South as usual produced slightly, but generally statistically significantly, higher mean scores on all the 'pencil and paper' tests compared with their counterparts in the North, the differences being of the order of two to three percentage points. The mean scores achieved by Midlands pupils either fall somewhere in between or are occasionally below the North.

Much greater differences have always been found when pupils were classified as attending schools drawing from distinctly different types of catchment area. On the basis of schools' responses about the proportions of their pupils they estimated were drawn from different kinds of catchment area, the schools were classified into

Table 4.2 *Pupils' subcategory performance levels in 1984 broken down by region in England*
(Mean percentage scores – weighted population estimates)

Subcategory		England	North	Midlands	South
Using graphs, tables and charts	mean s.e.	75.9 0.6	75.6 0.6	74.1 1.2	77.1 0.9
Making and interpreting observations	mean s.e.	50.1 0.5	50.7 0.9	45.4 0.8	49.2 0.7
Interpreting presented information	mean s.e.	44.1 0.7	42.3 1.1	43.6 1.4	45.7 1.0
Applying biology concepts	mean s.e.	32.5 0.5	31.3 0.9	32.9 1.0	33.2 0.8
Applying chemistry concepts	mean s.e.	32.0 0.6	31.1 1.1	30.2 1.1	33.7 0.9
Applying physics concepts	mean s.e.	31.3 0.6	29.8 0.9	30.2 1.1	33.1 0.9
Planning parts of investigations	mean s.e.	42.7 0.6	41.8 1.1	40.3 1.1	44.7 1.0

five catchment groups. As a general rule a school was classified into a particular group if its questionnaire response indicated that at least 80 per cent of its pupils were drawn from the respective catchment area. Where pupils were drawn more evenly from two or more different kinds of area the schools concerned were classed as 'mixed urban'. As Table 4.3 shows, there is a strong upward gradation in average pupil performance levels as the catchment area of comprehensive schools in England changes from inner city or less prosperous suburban, through mixed urban and rural to prosperous suburban.

Table 4.3 *Pupils' subcategory performance levels in 1984 broken down by catchment area in England*
(Unweighted mean percentage scores for pupils in English Comprehensives)

Subcategory	Prosp subur*	Rural	Mixed urban	Less prosp subur	Inner city
Using graphs, tables and charts	80	76	75	70	71
Making and interpreting observations	53	49	50	44	47
Interpreting presented information	48	46	43	37	39
Applying biology concepts	36	34	32	27	27
Applying chemistry concepts	36	32	32	27	25
Applying physics concepts	37	34	30	27	25
Planning parts of investigations	46	45	41	37	33

*Prosperous suburban.

This is a familiar performance pattern which has emerged consistently at every age in every science survey in which relevant information has been gathered (see Russell *et al*, 1988; Schofield *et al*, 1986). It has also been found to hold for Mathematics and for Language performance to similar degrees (see Foxman *et al*, 1985; Gorman *et al*, 1987).

A comment on the question of links between performance and school resources or subject take-up might be in order at this point. The kind of information described in Chapters 2 and 3 about the science experiences of pupils, and about the circumstances in which they undertake their science learning activity in school, is of interest in its own right. It is also essential information to have available against which to review pupils' science performances. This is particularly so if we consider any *changes* in these performance levels over time, either for the population as a whole or between pupil subgroups. For instance, should the well known performance gap between boys and girls in Physics appear to reduce over a period of time it will be essential to know if there have been changes in Physics take-up by boys *and* girls which might help explain the performance change.

Although of great interest to science educators and resource providers, relating performance to learning conditions more directly than this is an exercise fraught with difficulty. It is easy enough, of course, to explore statistical associations; the problems surround the interpretation of the results. Where an association is identified, as in the case of performance and laboratory supply in comprehensive schools noted in the 1980 survey report at this age (Driver *et al*, 1982), it can never be assumed to indicate a cause-effect relationship since there is always a confounding with one or more other variables. In this case, the comprehensive schools best provided with laboratories tend to be ex-Grammar schools in prosperous suburban areas.

This kind of investigation into links between pupil performance and school resources is not repeated here. What *is* included in this report, in Chapter 13, is an account of the findings which have emerged during an investigation into connections between performance and subject take-up. For sample size reasons such an investigation has necessarily waited until five years' of survey data have accumulated, and even here interpretation of the findings is subject to assumptions about pupils' performance levels *before* they made their option choices. In brief, the evidence is that courses in Chemistry and Physics, singly or, better, in combination, do seem to have a greater influence than others in developing the kinds of skills and abilities assessed in this survey programme. Readers are referred to Chapter 13 for details.

4.3 Reviewing performance patterns over time

Before reviewing the performance findings over the entire five year survey period 1980–84, it is essential to draw attention to three major factors which will have affected the reliability of performance estimates, and hence of subgroup comparisons, within and between surveys. It is particularly important to do this since the effect of these factors can sometimes only be guessed at and cannot be compensated for in analysis. The factors concerned relate to *question-sampling effects*, *school participation effects* and *marking effects*.

The analysis approach which has been adopted in this assessment programme is based on the assumption that the random samples of questions and of pupils (and their schools) involved in any survey represent their populations in known ways. The voluntary participation of schools in the surveys causes some problems in this respect. Not all selected schools accept the invitation to take part, and no information is available about the non-participating schools which would indicate whether they are different from those which do take part in any respect relevant to science. Table 4.4 shows the number of schools in England, Wales and Northern Ireland which have taken part in the surveys, and also indicates these as a percentage of those invited.

Table 4.4 *Numbers of schools which participated in the surveys*

(Bracketed figures indicate these schools as percentages of those *invited* to participate)

Year	England	Wales	Northern Ireland
1980	440 (80)	71 (81)	57 (85)
1981	336 (87)	109 (75)	104 (70)
1982	252 (83)	101 (69)	91 (64)
1983	363 (89)	112 (73)	115 (70)
1984	272 (82)	105 (73)	99 (68)

As Table 4.4 shows, the participation rates have differed from survey to survey in all three countries. In England the participation rate has varied from year to year at between 80 per cent and 90 per cent of invited schools. In Wales from 1981 onwards the participation rate has been between 70 per cent and 75 per cent, while in Northern Ireland the corresponding range is 65–70 per cent.

Schools which have declined to take part in particular years have sometimes done so because they had already been involved very recently in one or more previous annual surveys (in science *or* in mathematics *or* in language). The survey burden on schools in Wales and Northern Ireland has been especially severe since around half the secondary schools in *each* of these countries are invited to take part in any one APU survey. Indeed, in 1981 and 1982 when surveys were being carried out in all three curriculum areas at age 15 it was decided to select a single sample of schools in Wales and Northern Ireland and to administer Maths, Language *and* Science tests within each of these schools (different pupils taking different tests).

Whatever the reasons for schools declining to take part in surveys, the fluctuating participation rates must be a factor contributing to test score variation over the period. This is particularly so for Wales and Northern Ireland, given the relatively small numbers of schools which have represented these countries in each survey.

The second major but unquantifiable influence on test score variation over the *initial* survey period relates to the question-sampling aspect of test construction. As mentioned earlier, questions for survey administration are randomly selected from within pools of questions representing the various subcategories of the assessment framework. These random samples of questions are then assumed to reflect in appropriate ways the subcategory 'question populations'. A problem here is that these populations do not exist in practice in terms of all the suitable questions which could be classified into one or other of the subcategories. The pool itself is therefore a sample from this larger 'domain', and it will be more or less biased as it develops and grows in size – the extent of bias depending on the state of our knowledge of the nature of the domain concerned.

Cumulated testing experience has suggested numerous question features which appear to influence pupil performance; throwing doubt, incidentally, on the degree of 'purity' which can be assumed for the subcategory activities as these are described and assessed in the survey. Such features include, for example, mathematics dependence, strength of 'science context', form of data presentation, and so on. Clearly, it would be extremely difficult to attempt to construct a multidimensional grid which would adequately serve as a domain and hence pool and test specification. Indeed, such a task would have been impossible at the start of the survey programme when rather little was known about pupil performance in relation to the kind of assessment framework used here.

A consequence is that the comparability of the survey 'tests' which have been drawn from the growing pools from year to year in this developmental period must be in some doubt–in the earlier years because pool and test sizes were small and in later years because of the rationalization exercise mentioned in Chapter 1. For some subcategories the rationalization exercise resulted in pool restructuring, for others it resulted in the 'absorption' of large numbers of questions from age-pool into another. The developmental histories of the subcategories at age 15 are detailed later in this chapter.

The third unknown quantity is the influence of script marking effects. Marker reliability studies carried out during the 1980 survey indicated that there was high agreement between three independent markers in terms of the rank order in which they would place 300 or so scripts (based on the marks awarded). This is a traditional measure of marker 'reliability' but it is not completely appropriate in this context since severity/leniency is more important. When the marks awarded to the pupils by each marker were averaged there were differences of up to four percentage points in the three figures produced (see Driver *et al*, 1984, Appendix 8). It is usual practice in the surveys for two different markers to share the scripts relating to a particular test package – roughly 300 scripts each. Given this procedure there is no available method of adjusting for differences in the relative severity/leniency of different markers within or between surveys.

All three of the factors discussed above will undoubtedly have contributed to fluctuations in the performance level estimates over the five years, creating variation over and above that attributable to pupil, school and question sampling effects and to 'real' changes in the performance levels of the population of 15 year olds. It will also mean that differences will need to be rather substantial before they can be adjudged significant in any sense.

Bearing these comments in mind we move on to review the cumulated test data. Performance data are reviewed for five of the seven domain-sampled subcategories: **Using graphs, tables and charts, Interpreting presented information, Applying biology concepts, Applying chemistry concepts** and **Applying physics concepts**. The other two domain-sampled subcategories – **Making and interpreting observations** and **Planning parts of investigations** – have been assessed in only two surveys (1983 and 1984) in the same form, and so are not considered in this particular section. As mentioned earlier, the other two important practically based subcategories – **Use of apparatus and measuring instruments** and **Performance of investigations** – are not domain-sampled, but are represented in the surveys by small numbers of deliberately selected questions which are then reported in some detail. Again, therefore, these subcategories are not considered here; they are fully discussed in Chapters 6 and 11, respectively.

For ease of communication the mean percentage scores are rounded to the nearest percentage point. Comments are occasionally made about the statistical significance – at the usual five per cent level – of particular performance differences. As noted in Chapter 1, it should be recognised that statistical significance is not synonymous with educational significance. With large samples rather small differences often reach statistical significance when in practice they are of little educational importance. Similar it does not necessarily follow from a lack of statistical significance that a performance difference should be ignored. Differences which *persistently* emerge in the same direction in most or all surveys are likely to be worthy of attention and may well have pedagogical implications, whether or not they reach statistical significance in any one survey.

4.4 'Using graphs, tables and charts'

This subcategory – which is fully discussed in Chapter 5 – was created after the first survey by amalgamating two original subcategories: **Reading information from graphs, tables and charts** and **Representing information as graphs, tables and charts**. Any test package relating to this combined subcategory has contained an equal

Table 4.5 *'Using graphs, tables and charts' – developmental history 1980–84*

Survey year	No. of questions in pool	No. of questions in survey	Packaging strategy	Pupils/schools per package	Mean % scores Eng	Mean % scores Wales	Mean % scores NI
1980	93	60	4×15	750/180	61	55	57
1981	146	60	4×15	760/510	67	61	63
1982	146	64	4×16	800/420	61	58	58
1983	146	64	4×16	520/400	64	58	59
1984	216	90	6×15	440/400	76	73	75

number of randomly selected questions from each aspect – reading and representing. Table 4.5 provides the subcategory's developmental history.

In common with other similar tables in this chapter, Table 4.5 shows, for each year in which the subcategory was assessed: the number of questions available in the question pool; the number of questions randomly selected for survey administration; the way in which the questions were distributed among hour-long test packages; the approximate number of pupils who attempted each package and the approximate number of schools from which these pupils were drawn; and, finally, the estimated mean percentage scores for all 15 year olds in the three countries.

As with the other subcategories, we see from Table 4.5 that the number of questions available for selection in 1980 was relatively low, and that the pupil sample was highly 'clustered' (nine pupils in any school took the same package in 1980, compared with only one or two in later years). Between 1981 and 1983 pool size remained constant, and the test scores varied within sampling limits. Between 1983 and 1984 the pool increased substantially in size, and the test score jumped significantly also.

Both these features are attributable to the rationalization exercise referred to earlier, since in this case the majority of the 'new' questions actually originated in the corresponding pools at the other ages. These questions were always likely to be 'easier' for 15 year olds than those questions specifically written with this age group in mind, and this seems in fact to have been so. The problem of comparability has been further complicated by the fact that a number of mark schemes were altered to bring their philosophy into line with those in operation at the younger ages, for example, accurate point plotting in graph construction questions is now given more credit at this age than previously.

As far as the *reliability* performance scores for the pupils in the three countries are concerned the mean scores for England are *consistently* higher than those for Wales and Northern Ireland, significantly so in most surveys. At age 13 the pupils in England have also consistently produced higher scores than their peers in Wales, but at this lower age the performance scores of pupils in Northern Ireland have generally matched those of the English pupils.

Table 4.6 *'Using graphs, tables and charts' – the performance of boys and girls*
(Mean percentage scores – weighted population estimates)

Survey year	England Boys	England Girls	England Diff	Wales Boys	Wales Girls	Wales Diff	Northern Ireland Boys	Northern Ireland Girls	Northern Ireland Diff
1980	63	59	4	55	55	0	57	57	0
1981	68	65	3	63	58	5	61	65	−4
1982	62	60	2	60	57	3	59	57	2
1983	63	65	−2	57	60	−3	58	60	−2
1984	77	74	3	73	74	−1	74	76	−2

With the exception of 1983, there has been in England a consistent difference in favour of boys in the mean performance scores of 15 year olds in this subcategory. In 1980, 1981 and 1984 these differences have reached statistical significance. In Wales and Northern Ireland the picture is less clear with girls as often as boys taking the lead – a phenomenon apparent at age 13 in all three countries.

There is strong evidence in the performance data for individual questions that much of the overall difference in favour of boys can be attributed to their greater competence when handling coordinate graphs. As Chapter 5 reports, boys have produced higher mean scores than girls on about two-thirds of the graphical questions which have been used in at least one survey to date, whereas the girls have produced higher mean scores than boys on a similar proportion of questions featuring other kinds of representational forms. These relative strengths are present also in the survey data for 11 and 13 year olds (see Johnson and Murphy, 1986).

It is likely that the existing difference at ages 11 and 13 in the abilities of boys and girls when handling coordinate graphs might actually have been amplified after 13+ as a consequence of their differential Physics take-up rates. As Chapter 2 has shown, most boys but few girls choose to continue with Physics once this becomes optional. It has been suggested elsewhere (Johnson and Murphy, 1986) that the greater computational and graphical experience gained during Physics learning will therefore have given boys *as a group* this additional advantage. Indeed, the evidence of Chapter 13 would support this speculation.

Since the age 15 question pool has always contained a higher proportion of graphical questions than has the age 13 pool, this could explain the emergence of a consistent *overall* performance difference in favour of the

Table 4.7 *'Interpreting presented information' – developmental history 1980–84*

Survey year	No. of questions in pool	in survey	Packaging strategy	Pupils/schools per package	Mean % scores Eng	Wales	NI
1980	41	40	2×20	750/180	45	41	40
1981	63	54	3×18	760/510	41	36	41
1982	112	48	3×16	800/420	43	40	41
1983	129	64	4×16	560/430	41	37	37
1984	179	90	6×15	440/400	44	43	42

boys between these ages. Specifically, of the 140 different 'data representation' questions which have been administered in one or more of the five annual surveys at these ages, at age 13 one-third of these featured coordinate graphs compared with two-fifths at age 15.

4.5 'Interpreting presented information'

There is some overlap between this subcategory – fully described in Chapter 8 – and the one just discussed, since graphs, tables, charts and diagrams are used extensively to present the information which pupils are required to interpret. 'Interpretation' is here defined in terms of pupils' abilities to perceive systematic relationships between two variables; these are usually linear relationships if quantitative data are involved. Pupils are expected to demonstrate this ability in a variety of ways: they are sometimes, for instance, asked to describe the relationship, at other times to 'use' it by interpolating or extrapolating values, and so on. Pupils are *not* expected to bring anything but a very basic level of science knowledge or conceptual understanding to the problems. This is one of the ways in which this subcategory differs from those 'concept application' subcategories considered later.

As Table 4.7 shows, the question pool representing this subcategory has quadrupled in size since 1980, growing fairly steadily over the entire period. The proportion of questions selected for survey use in any year has decreased from virtually 100 per cent in 1980 to around half in 1984. In view of this the fluctuations in mean scores estimates over the period are to be expected. The fluctuations are arbitrary in direction and there is no suggestion in the data of any underlying trend in the true levels of pupil performance over the period.

The increase in pool size between 1983 and 1984 has again mainly resulted from the absorption of questions from the pools at the other ages. This strategy appears to have had a much smaller effect here than it did in the previous subcategory. One reason for this *might* be that there was a much lower degree of associated mark scheme change in this case.

Turning attention to the relative performances of pupils in the different countries, we see again that the pupils in England have consistently produced the higher scores – significantly higher in 1980, 1982 and 1983. As was the case for **Using graphs, tables and charts** the picture has changed for Northern Ireland in comparison with England between the ages of 13 and 15, since at the younger age (and also at age 11) the performances of pupils in these two countries have been similar. Survey evidence suggests that pupils in Wales have already lost parity with their peers in England by the age of 13.

The evidence in Table 4.8 is that, in England and Wales at least, boys tend consistently to produce higher mean scores than girls on the 'interpretation' tests at this age; in most years the performance differences reach statistical significance. This is particularly interesting in view of the fact that no such difference has emerged at age 11, and while performance differences have appeared at age 13 these have not been large and have not emerged consistently. Why then should boys be more successful than girls in this activity at age 15?

Table 4.8 *'Interpreting presented information' – the performances of boys and girls*
(Mean percentage scores – weighted population estimates)

Survey year	England Boys	Girls	Diff	Wales Boys	Girls	Diff	Northern Ireland Boys	Girls	Diff
1980	47	43	4	42	40	2	38	42	−4
1981	42	40	2	36	36	0	41	41	0
1982	45	41	4	41	39	2	41	41	0
1983	42	41	1	36	37	−1	38	35	3
1984	46	43	3	46	39	7	41	42	−1

It has been suggested elsewhere (Johnson and Murphy, 1986) that boys' superior graphical skills might be especially relevant in explaining this feature in the performance data. It has been noted earlier, and is discussed at greater length in Chapter 5, that boys seem to have a particular strength when handling coordinate graphs, and there have indeed been greater numbers of questions in the 'interpretation' tests at this age in which the data have been presented in graphical form. To be specific, at each age roughly 120 different questions have been administered in at least one of the five annual surveys, and of these six per cent, 15 per cent and 24 per cent of the questions featured coordinate graphs at ages 11, 13 and 15 respectively.

4.6 'Applying biology concepts'

This is one of the three 'concept application' subcategories included in the assessment framework at

Table 4.9 *'Applying biology concepts' – developmental history 1980–1984*

Survey year	No. of questions in pool	No. of questions in survey	Packaging strategy	Pupils/schools per package	Mean % scores Eng	Wales	NI
1980	50	48	2×20	750/180	42	39	40
1981	70	48	3×18	760/510	36	34	31
1982	109	48	3×16	800/420	33	32	30
1983	124	64	4×16	560/430	32	30	28
1984	124	66	6×15	440/400	33	32	31

age 15, and successful pupil performance in this particular case depends on an ability to apply pre-existing conceptual understanding in Biology (see Chapter 9).

Table 4.9 shows a now rather familiar picture of pool expansion. There were very few questions available for survey selection in this subcategory pool in the early years. Indeed, all the questions available in 1980 were administered in that first survey. It is highly likely that the apparent downward trend in test mean scores between 1980 and 1982 is a consequence of the gradual increase in pool size over this same period. Once the pool had reached 100+ questions Table 4.9 shows that the test scores began to fluctuate only very slightly and arbitrarily in direction – and certainly well within sampling limits.

There are only occasional instances of national performance differences reaching statistical significance in this subcategory. But a feature of some interest in Table 4.9 with regard to national comparisons is the *consistency* with which the pupils in Wales have performed slightly less well than the pupils from England, while the pupils from Northern Ireland have *consistently* produced the lowest mean scores (apart from 1980). This pattern of performance is particularly interesting since, as shown later, the pupils in Wales have usually produced identical performance scores to those of their peers in Northern Ireland when **Applying chemistry concepts** and when **Applying physics concepts** – both below the performance levels of pupils in England.

There is no evidence in Table 4.10 of any difference between the performance levels of boys and girls when applying biology concepts. This similarity holds even in Wales where boys consistently produced significantly higher mean scores than girls at age 13 (see Schofield *et al*, 1988). Where mean score differences *have* appeared these have usually been small and variable in direction, confirming previous reports for biology performance (see Comber and Keeves, 1973; NAEP, 1978a; Erickson and Erickson, 1984).

Table 4.10 *'Applying biology concepts' – the performances of boys and girls*
(Mean percentage scores – weighted population estimates)

Survey year	England Boys	England Girls	England Diff	Wales Boys	Wales Girls	Wales Diff	Northern Ireland Boys	Northern Ireland Girls	Northern Ireland Diff
1980	43	41	2	42	36	6	39	40	−1
1981	37	35	2	35	34	1	32	30	2
1982	33	32	1	32	32	0	30	30	0
1983	31	32	−1	29	30	−1	28	28	0
1984	33	33	0	31	33	−2	31	31	0

4.7 'Applying chemistry concepts'

As was the case for the *biology* 'concept application' subcategory, the question pool representing the subcategory **Applying chemistry concepts** was very small in 1980, only reaching its present size for the survey in 1983. It is hardly surprising, then, that the mean scores took a little time to settle down – again, any apparent 'trend' in the figures for England between 1980 and 1982 is more likely to be a consequence of this pool expansion than it is to indicate a 'real' underlying movement in the performance capabilities of 15 year olds. From 1982 onwards the performance scores fluctuate arbitrarily in direction and within sampling limits.

As Table 4.11 reveals, the pupils in England have more often than not produced slightly higher mean scores than their counterparts in Wales and Northern Ireland; the differences between England and both the other countries reach statistical significance in 1980, 1981 and 1983. The different take-up rates of Chemistry in the three countries must be a relevant factor in explaining England's small advantage here (see Chapters 2 and 13).

There is a general tendency revealed in the performance data of Table 4.12 for boys to produce higher mean scores than girls when applying chemistry concepts (the

Table 4.11 *'Applying chemistry concepts' – developmental history 1980–84*

Survey year	No. of questions in pool	No. of questions in survey	Packaging strategy	Pupils/schools per package	Mean % scores Eng	Wales	NI
1980	50	48	2×20	750/180	39	35	35
1981	65	48	3×18	760/510	37	33	33
1982	74	48	3×16	800/420	33	31	33
1983	104	64	4×16	560/430	31	29	28
1984	104	66	6×15	440/400	32	31	30

Table 4.12 *'Applying chemistry concepts' – the performances of boys and girls*
(Mean percentage scores – weighted population estimates)

Survey year	England Boys	England Girls	England Diff	Wales Boys	Wales Girls	Wales Diff	Northern Ireland Boys	Northern Ireland Girls	Northern Ireland Diff
1980	40	38	2	36	34	2	37	33	4
1981	39	35	4	33	32	1	33	33	0
1982	35	31	4	31	31	0	35	31	4
1983	33	30	3	30	28	2	28	28	0
1984	33	31	2	33	29	4	29	31	−2

performance differences in 1981 onwards reach statistical significance). This tendency is more strongly established in England than it is in Wales, and in both these countries it is more *firmly* established than it appears to be in Northern Ireland.

It is interesting that this performance difference when **Applying chemistry concepts** was not so well established at age 13 (see Schofield *et al*, 1988). What then can explain its emergence at the older age? Plausibly a link can again be suggested with the greater uptake of Chemistry by boys at 13+. Indeed, as Chapter 2 has shown, the difference in the percentage of boys and girls (in favour of boys) studying Chemistry at this age in *England* is larger than the corresponding difference in *Wales*, and this in turn is larger than the corresponding difference in take-up rates in *Northern Ireland*. It would be a natural consequence of this for the performance gaps between boys and girls in these countries to vary in sympathy–as they do.

4.8 'Applying physics concepts'

As Table 4.13 shows, the performance data for this subcategory over the entire survey period are some of the most stable so far discussed. Apart from the 1980 figure for England, the mean scores fluctuate within a span of four percentage points – well within sampling limits. Moreover, this stable picture has emerged *despite* the now familiar pattern of continuous growth of an initially very small question pool.

In general pupils in the three countries produce very similar mean scores on this subcategory, as Table 4.13 shows. Although the pupils in England have consistently produced slightly higher scores than those in Wales and Northern Ireland, they have not been significantly higher than *both* the other figures in every year (differences between England and *both* the other countries reach statistical significance in 1980 and 1983).

The relative strength and the consistency of the performance gap in favour of boys in this subcategory is clearly illustrated in Table 4.14. The size of the performance difference is of the order of half the standard deviation of raw scores in England and Wales. In Northern Ireland the differences are sometimes smaller than this; in these cases (1981 and 1984) it is the Northern Ireland *boys* who have produced lower mean scores than usual in comparison with their peers in the other countries. This is perhaps not a surprising feature in the survey data since the greater uptake of phsyics in England and Wales among 15 year olds is actually the result of greater physics taking by average and less able boys in these countries (presumably in turn a consequence of the comprehensive school systems operating here).

Table 4.14 *'Applying physics concepts' – the performance levels of boys and girls*
(Mean percentage scores – weighted population estimates)

Survey year	England Boys	England Girls	England Diff	Wales Boys	Wales Girls	Wales Diff	Northern Ireland Boys	Northern Ireland Girls	Northern Ireland Diff
1980	40	31	9	37	28	9	35	28	7
1981	35	27	8	33	27	6	32	28	4
1982	37	28	9	34	26	8	36	28	8
1983	35	28	7	32	25	7	32	25	7
1984	35	27	8	36	28	8	32	27	5

The Physics performance gap in favour of boys is, of course, well known. It has been reported previously to hold internationally (see Comber and Keeves, 1973) and, more recently, to hold in the USA (Hueftle *et al*, 1983) and elsewhere (see, for instance, Erickson and Erickson, 1984). It is also very firmly established by the earliest age surveyed (age 11 in the APU surveys, age nine in others), the only difference being an apparent increase between the ages of 13 and 15 (see Johnson and Murphy, 1986). Such an increase might reasonably be expected, of course, given the lower take-up rate of Physics by girls at 13+ (see Chapter 2).

Why such a performance difference *should* be present before children begin to experience much formal science education in schools has long been a puzzle. Lately, it has been suggested that clues might be found in children's early *out-of-school* hobbies and activities rather

Table 4.13 *'Applying physics concepts' – developmental history 1980–84*

Survey year	No. of questions in pool	No. of questions in survey	Packaging strategy	Pupils/schools per package	Mean % scores Eng	Mean % scores Wales	Mean % scores NI
1980	60	48	2×20	750/180	36	32	32
1981	81	48	3×18	760/510	31	30	30
1982	106	48	3×16	800/420	32	30	32
1983	129	64	4×16	560/430	31	28	28
1984	129	66	6×15	440/400	31	32	29

than in their in-school learning (Erickson and Erickson, 1984; Johnson and Murphy, 1986). Certainly the chosen activities of 11 year old boys and girls are indeed very different in general, the boys more often engaging in 'tinkering' activities involving mechanical or electrical objects and the girls more often engaging in biological or domestic activity (see Russell *et al*, 1988; Johnson and Murphy, 1986). This same dichotomy in interests is continued into their teens, as Chapter 3 has shown.

4.9 Summary

There is no evidence in the survey data of any underlying trend in the absolute subcategory performance levels of 15 year olds between 1980 and 1984. Any systematic changes in performance score estimates for this period can be linked to corresponding changes in question pool and test sizes. Nor is there any evidence of changes in the size of performance gaps between particular pupil groups – such as those in the three countries, or of different sexes, The picture is a relatively stable one, as far as can be ascertained from the test data.

Performance levels for **Using graphs, tables and charts** have always been higher and performance levels on the 'concept application' subcategories have always been lower than have those of all other subcategories. Moreover, this relative pattern holds for ages 11 and 13 also. Unfortunately, such comparisons are never readily interpretable. For example, it is not necessarily the case that the 'skill' of data representation is inherently 'easier' or even better practised than is that of concept application (although both are likely). Pupils are merely more successful in demonstrating these different skills given the kinds of questions used to assess them in these surveys, and given the particular mark schemes applied.

Whether this profile of subcategory performance can be modified by a different emphasis in teaching is open to debate, although there is some evidence available from these science surveys which suggests that different science courses do affect performance in the various subcategories to differential extents. This issue is discussed in Chapter 13.

A notable feature in national performance comparisons is the consistency with which pupils in Wales and Northern Ireland fail to match the performance levels of their peers in England in some subcategories, when their performance levels at age 11 were very similar. This can for the most part be attributed to the differential rates of science uptake at 13+ in these three countries, given that a greater proportion of 15 year olds study at least one science in England than in the other countries. The only problem with this interpretation is that the slightly lower average performance levels of 15 year olds in Wales emerge earlier – at age 13.

Turning attention to gender related differences in performance, boys significantly outperformed girls in every survey (and at every age) when **Applying physics concepts**, and in most surveys (at age 15 only) when **Using graphs, tables and charts, Interpreting presented information** and **Applying chemistry concepts**. Girls have usually outperformed boys on average when **Making and interpreting observations**. For **Applying biology concepts** and **Planning parts of investigations** performance differences have always been small. However, as Chapter 13 shows, when the differential course uptake of boys and girls is taken into account all performance gaps *except* those for **Applying physics concepts** and **Making and interpreting observations** disappear. These performance gaps persist to the same extent independently of the recent science course experience of the pupils; even the more able girls still studying Physics fail to match the group mean performance score of the boys. This suggests a strong early established performance difference, plausibly arising from differences in the science relevant out-of-school activities and interests of boys and girls as young children.

5

Using graphical and symbolic representation

5.1 Introduction

The ability to use graphs, tables and charts of various kinds is important within and beyond work in science. In everyday living people are constantly faced with the need to read information given in these various forms. Bus and railway timetables are perhaps the most familiar examples of tabular displays of information which everyone meets in the course of daily life. Graphs and charts, on the other hand, are rather favoured forms of data presentation in television news programmes. While the *compilation* of tables and the *construction* of graphs and charts are not such widely needed skills in this broader context, they are very much in evidence in science and are also important in a number of other subject areas. Indeed, graphical work is a popular component in both science and mathematics activity in primary as well as in secondary schools, and data representation skills in general are important in other subjects in the secondary school such as Economics, Statistics and Geography. Pupils' abilities to handle information in these various forms are here assessed in the subcategory **Using graphs, tables and charts**.

The questions which have been developed for use in the Science surveys are very varied in style and content. The main forms of data representation are naturally included: tables, bar charts, pie charts, coordinate graphs and grids. Also present are questions featuring flow charts, Venn diagrams, food webs, area plans, and so on. Thus, the more 'modern' forms have a place alongside the traditional ones, although the latter forms predominate. Where the information in the questions takes the form of statistical data this can be of various kinds from a variety of contexts. Not surprisingly, 'experimental results' in science are in evidence, but so also are economic data such as energy consumption or tourism figures, and demographic statistics including housing survey data and road accident statistics amongst other kinds.

The subcategory **Using graphs, tables and charts** is one of the 'domain-sampled' subcategories in the assessment framework. As the previous chapter has shown, the size of its question pool has gradually increased until by 1984 over 200 suitable questions were available for survey selection. Typically 60 questions are randomly selected for use in any survey, and over the period the total number of *different* questions which have been administered at least once is 141.

Within science, of course, there is a need to move beyond these generally applicable kinds of representational forms and to be additionally familiar with other *science specific* symbolic representations. For instance, the language of Chemistry is based on chemical symbols and equations; in Physics there are conventional ways of representing electrical circuits as two dimensional circuit diagrams; and in all the sciences there are conventions for representing apparatus set-ups as section drawings. Questions intended to assess pupils' abilities to handle these very specific forms are included in the subcategory **Using scientific symbols and conventions**.

The symbolism concerned is *highly* specific to one or other of the three main sciences and is relatively limited in variety. In view of this it was decided not to attempt to create large numbers of questions to be domain sampled each time, but rather to administer just a handful of specially selected questions to provide information about pupils' familiarity with particular types of symbolism. Thus there might in a single survey be just three questions from this subcategory: one featuring a circuit diagram, another asking pupils to label the pieces of apparatus in a section drawing, and perhaps a third looking for evidence of understanding of chemical formulae.

This chapter begins with the main subcategory **Using graphs, tables and charts**, looking in turn at the various different representational forms and summarising our cumulated knowledge of pupils' capabilities with respect to these. A brief section is devoted to the evidence of pupils' familiarity with the science specific symbolism just discussed. Some general points will be drawn together towards the end of the chapter; in particular comments are offered on the pattern of national and gender related differences in performance.

In commenting in this chapter on the relative performances of pupils of different abilities, reference is made to pupils in the 'top performance band' and those in the 'bottom performance band'. The top band contains the highest scoring fifth of pupils and the bottom band the lowest scoring fifth in terms of overall test scores.

5.2 Reading and inserting into tables

In principle both reading *and* compiling tables are activities embraced by this subcategory. *In practice* the majority of questions featuring tables require pupils to show their ability to *read* tabulated information. It might after all be a rather artificial exercise to present pupils with a large set of unordered data to be reorganised and presented in tabular form – particularly in pencil and paper tests. Some questions do exist in the 'data representation' question pool which require pupils to organise arbitrarily listed data values into tabular form, but they are very few in number. This would be a more appropriate activity to assess during a practical investigation, perhaps, where pupils would gather their own data and tabulate in the process – some of the practical questions in other subcategories have indeed afforded pupils this kind of opportunity, as discussed later.

Questions generally present data already in tabular form and require pupils to read information from these tables, to insert isolated values appropriately into them, or to transform the same data into another form – usually a bar chart or coordinate graph. The first kind of question, of which 12 have been administered in at least one survey to date, is specifically intended to explore pupils' abilities to read information from tables. The second kind of question, of which nine have been used, represents an attempt to assess in a rather marginal way pupils' abilities to compile tables. In the third kind of question the focus of assessment is the adequacy of the pupil's attempt at presenting the given tabulated data in the alternative form – the ability to read the data from the presented tables must here be assumed and is not credited in the mark schemes (pupil performance on these questions is discussed in following sections).

Given the cumulated survey data, what can now be said about pupils' familiarity with tables? Unfortunately, as with all the other kinds of representational forms to be discussed later – indeed as in every aspect of the entire assessment framework in science *and* other curriculum areas, it is not possible to offer *unqualified* generalizations about pupils' capabilities. It *can* be said that 90 per cent or more of 15 year olds are able to read and to insert individual values appropriately, provided the tables are simple and uncluttered, ie provided they do not contain decimal numbers and provided no arithmetic operations are required. Success rates decrease as table complexity increases.

An example question presented and discussed in the report of the 1980 survey (Driver *et al*, 1982, p26) will usefully serve to illustrate this general pattern. The table in this question was concerned with a comparison of death rates in car accidents in seven European countries in 1966. For each country, the total number of deaths was recorded, along with the number of cars on the road (in millions) and the number of deaths per million cars (decimal numbers here). Pupils were to indicate the number of deaths there had been in that year in one of the named countries, to indicate the number of cars there were on the road in another named country, to name the country with the fewest cars on the road, and to name the country with the smallest number of deaths per car. Just about a third of all the pupils who attempted this question gained all four available marks. The contrast between the most and least able pupils is quite stark: 83 per cent of the pupils in the top performance band gained all four marks compared with just one per cent of those in the bottom band, in practice a single pupil! A full quarter of the least able pupils failed to gain any marks at all.

The evidence from the 21 questions which are available with performance data is that the majority of the pupils in the top performance band can cope very successfully and consistently with tables. These pupils, as a group, achieved mean scores of 90 per cent or more on most of these questions; 95 per cent, for instance, on the question just described. Indeed for six of the nine 'insertion' questions these most able pupils produced mean scores of 100 per cent. Among the pupils in the bottom performance band the picture is rather different. Question mean scores for this group have varied between four per cent and 96 per cent; 30 per cent on the question above. The highest score was achieved by the least able pupils as a group for a question in which weather symbols had to be inserted appropriately into a 'weather chart', ie a table of symbols indicating the weather day by day. Among the least able pupils the lowest scores tended to be associated with those questions featuring tables of numerical data.

As mentioned earlier, there are one or two practical activities from other parts of the assessment framework which afford pupils the opportunity to gather data and to record this in an appropriate form. In one such question, from the subcategory **Use of apparatus and measuring instruments**, pupils were presented with a beaker of boiling salt solution, and were instructed to record the temperature of the liquid over a three minute period as it cooled. They were to present their results in tabular form. Around 80 per cent of the tables produced by the pupils were judged to have been presented in a readable fashion, showing discrete columns of figures for temperature and time. Imposition of more stringent criteria indicated that just over 40 per cent of the pupils' tables were well presented and coherent. About one in five pupils labelled the tables with physical quantities *and* appropriate units for both variables.

Some of the longer practical problem solving tasks from the subcategory **Performing entire investigations** (see Chapter 11) have provided the opportunity to see how pupils in practice, and *without* direction, set about the task of recording observations. One such question asks pupils to determine which of three kinds of paper towel is most absorbent, another asks pupils to find out how the angle of a taper affects the rate of burning. The two questions were administered to different samples of pupils in 1981–82. In responding to these questions the pupils needed to make and record a number of observations as they set about solving the problems. It is of some interest to find that while in total 70–90 per cent of the 15 year olds tested presented their data in a systematic way, the proportions who actually compiled a formal table of results were 13 per cent in the first question and 28 per cent in the second (see the report of the 1982 survey – Gott *et al*, 1985, Chapter 10). Thus, while it is the case that the majority of 15 year olds are able to handle straightforward tables quite competently when this is the task in focus, perhaps too few readily apply this ability voluntarily in new situations.

5.3 Reading and constructing bar charts and histograms

Just 22 bar chart or histogram questions have been administered in at least one of the annual surveys–seven of these have required pupils to read information from these forms, 12 have presented pupils with partially completed charts to which they have been asked to add additional given data, and three questions have required them to produce charts or histograms from scratch to present given information. The majority of these questions have been concerned with illustrating frequency distributions of various kinds.

The evidence from the performance data is that the majority of pupils feel comfortable about handling these forms. The simplest tasks within any of the questions have required pupils merely to identify the longest or shortest bars in order to produce a successful response. An example would be a pictograph showing the numbers of ice creams sold each day of a week, in which pupils are asked to indicate the day on which the most ice creams were sold. The evidence is that more than 90 per cent of 15 year olds can do this. Almost all the pupils who failed in this task were in the bottom performance band.

Whenever there has been a need for further manipulation of presented data pupil performance has fallen markedly. The requirement to process information spanned by more than one bar has resulted in a variety of errors, suggesting either a failure to appreciate the meaning of a bar interval or an inability to effect the necessary computations accurately (typically pupils need to add several readings or to subtract two). Grouped data have proved particularly troublesome in this respect. An example question which nicely illustrates the problem is one which was presented and discussed in the report of the 1981 survey (Driver *et al*, 1984, p61). This featured a histogram showing the numbers of woodlice found in areas of different humidity half a day after being placed together in a perspex container (science teachers will recognise this popular 'experiment' which recurs in various forms throughout the assessment framework). Each bar interval spanned a range of 10 percentage points of humidity, and the pupils' task was to indicate how many woodlice were found in areas whose humidity was within the range 40 per cent to 69 per cent.

To answer the question successfully the pupils needed to take three consecutive bar readings and to sum the three woodlice counts. Just half the 15 year olds tested managed this task.

As indicated earlier this rather high degree of failure among the less able pupils suggests *either* a lack of understanding of the meaning of grouped data intervals *or* an inability to effect the necessary computations accurately. The APU Mathematics survey findings certainly indicate a lack of understanding among even these older pupils of the principle underlying the simplest frequency block graph. Pupils were presented with a block graph showing the numbers of houses in a street containing particular numbers of people, one 'block' per house, and inhabitants numbering between one and four. They were asked for the number of people living in the street. Just two-thirds of the pupils tested gave the correct response. More than one in five of these 15 year olds simply added the bar labels, ie they added the numbers on the horizontal axis (see Foxman *et al*, 1985, p394).

To test pupils' ability to *construct* bar charts, two general kinds of question have been used in the Science surveys. In the first kind pupils have been presented with partially completed bar charts (axes drawn and fully labelled) to which they were required to add extra data. Twelve such questions are available with performance data. In the second kind pupils have been presented with a modest amount of tabulated data and a fresh piece of 2 mm pitch graph paper, and have been asked to present the data in the form of a bar chart or histogram, as appropriate. Just three such questions are available with performance data.

On average 80 per cent or more of the pupils could insert given information into a pre-existing bar chart successfully. An example question was presented and discussed in the report of the 1981 survey (Driver *et al*, 1984, p65). This question happens to feature a bar line chart showing the numbers of pods containing particular numbers of peas ranging from 0 to 10. Three bar lines are already drawn, and the pupils are to draw in three more to show additional data. Three-quarters of the 15 year olds were considered completely successful with all three insertions; for each individual bar line 80–90 per cent of the pupils produced adequate attempts. These are typical results for this kind of question.

Turning attention to pupils' abilities to construct a bar chart or histogram from scratch, the evidence from the science surveys is that only a minority of 15 year olds can produce an appropriate graph complete in every sense. One question, again presented and discussed in the report of the 1981 survey (Driver *et al*, 1984, p66), serves to illustrate the general points. In this question pupils were presented with a table of the 'results' of a scientific investigation into the effects of a fertilizer on seedling growth. The table recorded the numbers of seedling leaves (from a total of 30) whose widths were within particular consecutive 3 mm ranges. The pupils were asked to draw a histogram to illustrate this data. About 70 per cent of the pupils chose sensible scales for the axes, using more than half the width or height of the paper, and labelled the axes appropriately; about 60 per cent accurately drew the bars. Fewer than a quarter of the pupils tested were able to construct a complete and appropriate graph successfully.

The APU Mathematics surveys confirm these findings: in these surveys about 85 per cent of 15 year olds were found to be successful at reading block graphs and bar charts given a simple scale, while 75–85 per cent were able to draw such graphs given some degree of help within the question (see Foxman *et al*, 1985, p388). Again, as illustrated above, performance levels dropped with the introduction of any greater complexity into the questions. (No questions have been used in the Mathematics surveys which required pupils to produce a graph from scratch.)

The polarization between the most and least able pupils is again marked. The evidence is that all but a handful of the pupils in the top performance band can read and produce bar charts successfully and consistently. In contrast, the least able pupils are not only less successful in every aspect but, as a group, are quite inconsistent from one question to another. A brief examination of the percentage mean scores achieved by these two groups on the bar chart and histogram questions will suffice to illustrate the difference between them in this respect. The mean scores across all 22 questions range between 85 per cent and 100 per cent for the pupils in the top performance band. For the pupils in the bottom performance band the mean scores range between six per cent and 95 per cent. Typically around 40 per cent of the least able pupils have scored zero for their construction efforts.

Since all the histograms presented in the science questions are equal-interval it is not possible to comment on the extent to which 15 year olds, of any ability level, understand the implications of the area under the graph in this context.

5.4 Reading and constructing coordinate graphs

In total, 47 graphical questions are currently available with pupil performance results; 28 of these present line graphs from which pupils are to read information as directed, just four present a partially completed graph to which pupils are to add given data points, and 15 require pupils to construct a graph from scratch given a table of data. About a quarter of the 'reading' questions are particularly complex in presenting superimposed graphs; half the questions require some arithmetical manipulation of readings to provide an appropriate response; about a quarter demand a degree of gradient interpretation. In all these ways the graph questions presented to 15 year olds differ from those presented to 11 year olds or to 13 year olds in these Science surveys.

Almost without exception the graphs features in the questions are temporal records of some kind, and most would be seen as related more to the physical sciences than to the biological sciences or to everyday applications. For instance, 90 per cent or so of the questions feature distance-time or speed-time graphs, cooling curves or solubility graphs. A handful of questions have a biological context by featuring population curves or growth curves. This 'content' characteristic is especially important to note since it is in contrast to the prevalence of sociological or economic data in the questions featuring other representational forms.

The survey evidence is that about 80–90 per cent of 15 year olds can successfully provide the second coordinate of a point given the first coordinate, or can insert individual points into a graph, provided the coordinates of the points coincide with *marked* divisions on the axes. Indeed, other researchers' findings have provided similar evidence of pupils' relatively high level of ability in this rather simple task (see, for instance, the CSMS project's report on graphs – Kerslake, 1981, p120). When one coordinate falls between labelled scale gradations the success rate drops to around 70 per cent of pupils; when *both* coordinates fall between labelled scale gradations the success rate drops to around 50–60 per cent of pupils.

The problem which many pupils experience when attempting to interpolate simple scales, or to read complex scales at all, appears to be quite general, since it has emerged also during instrument reading. In these Science surveys, when pupils have been asked to read various measuring instruments set at pre-determined values in the practical category **Use of apparatus and measuring instruments** they have usually coped well when the set value has coincided with a marked scale division on a simple scale, but have often floundered when the value lay between marked points (see the next chapter). The same phenomenon has been demonstrated to an equivalent degree in pencil and paper 'scale reading' questions in the APU Mathematics surveys, and also in graph questions (see Cambridge Institute of Education, 1985, Chapter 7). The more complex the scale – marked divisions in counts of two, for instance – the more difficulty many pupils experience.

In the 'graph construction' questions pupils have usually been presented with a modest table of data, typically six or seven pairs of readings, and a blank sheet of 2 mm pitch graph paper and have been asked to plot an appropriate graph. Usually, about a fifth of the tested pupils have not attempted a response. The pupils are not asked specifically to draw a *line* graph, so it is of some interest that, of those who did respond, at least 95 per cent chose to produce such a graph. The only exception to this is a question showing plant height readings over a five week period; the *majority* of pupils attempted this question and an unusually high proportion (a quarter) chose to present a bar chart – perhaps suggesting the influence of their early school experience, since plant growth is a favourite topic for bar chart work in the primary school.

Two questions were presented and discussed in the report of the 1980 survey (Driver *et al*, 1982, p35–36), and the results still serve to illustrate the general pattern of response for all the graph construction/completion questions. The first is an example where fully labelled axes are provided and the pupils are required to plot points and draw an appropriate line. A table is given containing six pairs of readings – claimed this time to be the observations made during a chemical reaction, in which the amount of residue was measured every two hours over a period of 10 hours. The second question presented pupils with a table of eight readings relating blood output (litres/minute) to heart rate (beats per minute). The intervals between the heart rates at which measurements were given were unequal.

The cumulated findings from these and similar questions is that, *in general*, when constructing a coordinate graph from scratch, about two-thirds of 15 year olds set up sensible (defined as using more than half the page) equal-interval scales for both axes. This finding is in line with that reported for 15 year olds in the CSMS project – see Kerslake, 1981. About half these pupils labelled both axes adequately, with variable names and appropriate units given. About half of all pupils plotted most or all the points to within half a scale division. A quarter to a third connected every pair of neighbouring points with straight line segments, while a similar proportion drew a smooth curve through most or all the points. Just about one in ten of all these 15 year olds produced a 'line of best fit'. It is interesting that this last figure coincides with that reported for a sample of pupils at this age who plotted their own measurements of length against time for a burning taper during the practical problem solving task described earlier – Gott *et al*, 1985, p204.

Once again there are marked differences between the pupils in the top and bottom performance bands in terms both of the level and the consistency of their performances. The pupils in the top band have, on average, achieved percentage mean scores on these questions ranging between 70–100 per cent (most questions above 90 per cent) for both 'reading' and 'constructing'. In contrast, the pupils in the bottom band have as a group produced percentage mean scores between four per cent and 80 per cent for the 'reading' questions and between one per cent and 50 per cent for the 'constructing' questions.

5.5 Reading and completing pie charts

Over the period of the annual surveys 15 different pie chart questions have become available with pupil performance data; seven of these present pupils with a complete chart from which they are to read information as directed, and eight present partially completed charts – ie pre-sectored with or without some existing shading – and require pupils to insert given information appropriately. For the most part these questions present economic or demographic data, usually in the form of frequency distributions resulting from surveys of various kinds, for example housing, modes of travel, food preferences, investment breakdown. The topic of nutrition provides another useful source of appropriate data in the form of food constituents, eg proportional composition of egg yolk in terms of major ingredients.

The evidence of the surveys is that a high proportion of 15 year olds are able to handle pie charts successfully when these involve a small number of simple sectors. Typically, 90 per cent or more of 15 year olds can successfully read information from and can add information to pie charts where this involves identifying a major or minor component in the whole, or counting or shading complete sectors. Anything more complex depresses general performance levels.

Two example questions were presented and discussed in the report of the 1980 survey (Driver *et al*, 1982, p28 and p30). Both these questions presented pupils with a modest table of data and an unshaded pie chart marked into ten equal sectors. The data given in one of the questions purported to be the results of a travel survey in which 100 people were asked how they travelled between London and Glasgow (rail, air, bus, car). The given artificial data is the simplest possible in that all the 'counts' in the table are multiples of ten – with the consequence that they map neatly onto multiples of marked sectors. A high 99 per cent of the pupils in the top performance band scored full marks for their responses, and even among the pupils in the bottom band more than half were able to deal with this question entirely successfully.

Once again, though, it seems that when interpolation of marked divisions is needed success rates fall among the weaker pupils. The second of the two example questions serves as an appropriate illustration. The data presented in this question were claimed to be the results of a housing survey conducted in overcrowded areas of London in the last century. This time the numbers given are not raw counts but percentages (of population living one, two, three or more persons to a room), and this time they do not exactly map onto the marked sectors but involve half sectors in two cases. While this caused the more able pupils no problems (96 per cent scored full marks) it does seem to have thrown the less able pupils somewhat. Just one in five of the pupils in the bottom performance band scored full marks this time.

In general, it can be said that the more able pupils experienced no difficulties with any of the pie chart questions. Their group mean scores for the individual questions were consistently at or approaching 100 per cent. Consistency in performance weakened as usual with a move down the five performance bands; among the pupils in the bottom band the question mean scores varied between 20 per cent and 83 per cent (this latter a food constituents question).

5.6 Reading and completing flow charts, food webs, Venn diagrams

In both Mathematics and Science *flow charts* are used to simplify the expression of complex and usually dynamic information. In Science this representational form is frequently used to illustrate the stages in a life cycle. Industrial processes (technological applications) are also usefully described in this way. Just six flow chart questions have been administered in the Science surveys over the five year period. Three have presented pupils with completely drawn and labelled charts and have required them to read information as directed from these; the other three have presented completely drawn but only partially labelled charts and have asked the pupils to label stages appropriately given a prose description of the process or relationship portrayed.

Given so few examples, it is not easy to generalize about pupils' abilities to handle this form of information representation. However, as with the other forms the pupils in the top performance band handled these questions rather well – they produced mean scores above 90 per cent for five of the six questions. Those pupils in the bottom group were again inconsistent, with question mean scores ranging between 17 per cent and 87 per cent. The only question which posed the more able pupils some problems was one which illustrated the stages in an industrial oil refining process. This question differed from the others both in terms of the density and difficulty of the information presented and in terms of the complexity of the chart itself – chemicals added into the process at different stages, the chart resembling a section drawing of apparatus rather than the more usual 'box and diamond' form. The chart which even the least able pupils found relatively easy to handle was concerned with an 'everyday' application – computer booking of airline tickets – and presented a simple chart with a small number of 'action boxes' and two or three 'decision points'.

An example 'insertion' question was presented and discussed in the report of the 1980 survey (Driver *et al*, 1982, p27). This question provided a prose description of the life cycle of a moss plant accompanied by an illustrative and partially labelled flow chart. Pupils were asked to name four unlabelled stages. Only 16 per cent of the pupils who attempted this question were completely successful, and these were for the most part those in the top performance bands. Half the pupils in the top band scored full marks, and all but a handful gained at least half marks. In contrast, none of the pupils in the lowest performance band was totally successful, and a high 70 per cent of these pupils gained no marks at all.

The successful completion or interpretation of *food chains* and *food webs* relies mainly on an understanding of the conventions used but also depends on the complexities of the biological relationships so represented. Although again just six relevant questions are available with performance data, these suggest that the majority of 15 year olds (more than 85 per cent) have demonstrated an ability to use correctly the arrows which signify the flow of energy in a food chain, and to interpret such a representation appropriately. Where pupils have failed in these questions, they have usually succumbed to a common misunderstanding and have attached the wrong directional meaning to the arrows. It will come as no surprise that the majority of those pupils who have shown evidence of a lack of knowledge of the arrow convention, or who have simply failed to apply this knowledge appropriately, have in general been among the least able. Very few of the average or most able pupils found any difficulty in using this particular form of representation.

Just four questions featuring *Venn diagrams* are available with performance data. The two in which pupils were required to read information from a Venn diagram were both concerned with animal characteristics. One of these questions was actually presented and discussed in the report of the 1981 survey of 11 year olds (Harlen *et al*, 1983, p161). The diagram showed three intersecting sets; set membership was defined, respectively, as 'mammalian', 'eats grass' and 'has four legs'. The task was to describe an animal – OKAPI – situated in the intersection of all three sets. All the 15 year olds in the top performance band and the majority of those in the next two bands scored full marks. Even among the least able pupils most gained marks for this question. Indeed, among the 11 year olds tested a high 70 per cent produced an entirely successful performance on this question. The performance pattern for the other related question was very similar. Putting information *into* Venn diagrams proved an equally straightforward exercise for the most able pupils, but a rather more difficult task for the rest.

5.7 Using chemical equations, circuit diagrams and section drawings

The available survey findings about pupils' use and interpretation of *chemical symbols* are patchy. This is simply because, while there are many different symbols in use, and different elements, compounds or mixtures about which questions might be asked, relatively few questions have actually been used in the surveys to represent this particular activity. As mentioned earlier, there might only be two or three questions administered in any one survey to represent the entire subcategory **Using scientific symbols and conventions**. Indeed, none was administered in the surveys after that of 1981.

One relevant question which was administered in the 1980 survey was intended to explore pupils' familiarity with the conventional symbol for chemical equilibrium. Given a series of alternative signs and asked which represents the equilibrium state in chemical formulae, half the sample pupils chose the correct option. Another question, given to a different sample of pupils, presented pupils with the equilibrium symbol and asked them to explain in full what the symbol meant. About a third of the pupils responded successfully. Among the pupils in the top performance band, two in three identified the symbol as denoting chemical equilibrium; in contrast, fewer than one in twenty of the pupils in the bottom band did so. Among those more able pupils who were currently studying Chemistry, almost all gave the correct meaning. This is an expected result, of course, given that these questions are entirely concerned with tapping pupils' knowledge of learned conventions, and it is unlikely that many would meet chemical symbolism outside Chemistry classes.

Turning now to pupils' familiarity with the conventions used to represent electrical components and circuits in formal *circuit diagrams*, the evidence is that 50–60 per cent of 15 year olds are familiar with the conventional symbols for batteries (cells), switches and bulbs. About 45 per cent can successfully identify the conventional symbols for ammeters and voltmeters. Complementary findings have emerged during practical testing for the category **Use of apparatus and measuring instruments** which is described in the next chapter. When faced with a limited choice of components in a practical situation, in preparation for assembling a circuit, between 80 per cent and 90 per cent of pupils have made correct selections of bulbs, cells and switches following a circuit diagram. The degree of success has usually fallen when meters were involved, with 50–60 per cent of pupils selecting the appropriate meter.

Following a circuit diagram to connect components into a circuit, or representing a given circuit in conventional diagrammatic form, might be seen as direct applications of this basic knowledge. However, a relatively low 30–40 per cent of pupils used the accepted symbols for meters and other components when translating a pictorial representation of a circuit into a formal circuit diagram. While 33 per cent of the pupils produced a diagram which matched the presented drawing in essential structural features, only 17 per cent of the pupils did so with *all* the individual symbols correctly used (see the report of the 1980 survey, Driver *et al*, 1982, p40). In a practical question – again from the category **Use of apparatus and measuring instruments** – in which pupils were to follow a formal circuit diagram to set-up a working circuit, about 60 per cent connected cells and switches, 20–35 per cent joined the bulbs into the circuit as indicated, but only 14 per cent produced a completely correct circuit.

The subject specific nature of circuit diagrams is unquestioned. The Physics takers in the pupil samples were far more successful in all the activities described above. They showed more than twice the success rate in both following and producing circuit diagrams. In contrast, when pupils have been asked to connect a working circuit by following a *photograph* of the circuit rather than a circuit diagram 70 per cent of them were successful, and there was little difference in success rates between Physics takers and the rest (see Gott, 1984).

As far as *sectional drawings* are concerned, the evidence is that most 15 year olds would be able to follow a sectional drawing to set up apparatus correctly, despite the fact that most of these pupils would be unable to name all the individual items in such a drawing or produce a similar drawing accurately themselves. When asked simply to name indicated pieces of equipment in a section drawing of the apparatus set up for a gas collection procedure, rather few pupils managed the task – for instance, 15 per cent named the delivery tube, 17 per cent the flat bottomed flask and 43 per cent the bung. Given a three dimensional drawing of a gas identification set-up involving two test tubes and asked to produce a corresponding section drawing, about a fifth of the pupils tested actually produced a *true* section drawing correct in most or all respects. This figure corresponds rather closely with that just given previously for drawing circuit diagrams. The main errors made were most usually of detail – liquid level omitted, bungs drawn in three dimensions, and so on. Detailed results for both these questions are given in the report of the 1980 survey (Driver *et al*, 1982, pp41–2).

Complementary findings are available once again from the practical testing of the category **Use of apparatus and measuring instruments**, since one question required pupils to set up the gas delivery equipment illustrated in a section drawing, given a suitable choice of items of equipment. Well over three-quarters of the 15 year olds tested followed the section drawing successfully, by choosing the correct items of equipment and setting them up appropriately. This high success rate is in line with that for the question described earlier in which pupils were to reproduce an electrical circuit shown in a photograph – a not unexpected result, perhaps, given that the representations of equipment in section drawings are rather close to their three dimensional realisations.

5.8 Ability-related, gender-related and national differences in performance

Throughout this chapter comments have been made about the relative capabilities of pupils in the top and bottom performance bands, ie pupils in the top 20 per cent and bottom 20 per cent of the ability range in terms of overall test scores. In brief, the evidence is that among the most able group few pupils experienced any difficulty at all with the kinds of questions discussed in this chapter. Indeed, most pupils in the top 40 per cent of the ability range were successful on the majority of questions. Among the lower ability groups success rates fell as the tasks became more complex. Scale reading proved troublesome for many pupils, readings between marked scale gradations provided difficulties as did scales not directly representing counts of 10, 100, 1,000, etc. Thus, scales marked in twos, say, or marked in units with each unit representing a multiple of some kind (as, for example, in typical cartographic map scales) proved problematic. Arithmetic manipulations posed further difficulties for the less able pupils.

Individual questions usually incorporate a series of often quite different demands, and it is interesting to see the pattern of success rates for whole questions for the five performance bands. This information is given in Table 5.1, which shows for each performance band the percentage of questions with overall mean scores within certain limits. The table shows that the pupils in the top group scored over 80 per cent on 86 per cent of the questions. Indeed, for 70 per cent of the questions the mean scores for this group were at or above 90 per cent. This reflects the situation already described, in that few of the most able pupils found any difficulty with the questions set. Among the least able group the picture is less clear, since the questions are distributed fairly evenly over the bottom three – even four – score ranges.

As far as gender differences in performance are concerned, there has consistently been a statistically significant difference in overall test scores in favour of boys at this age for the subcategory **Using graphs, tables and charts** (see previous chapter). This overall difference masks differences between boys and girls in terms of their relative success with different forms of representation. In brief, there is evidence that boys are on the whole slightly more successful than girls when handling coordinate graphs while girls are on average slightly better than boys at dealing with other representational forms.

Table 5.1 *Percentage of questions at different difficulty levels for the pupils in five performance bands*
(141 questions in total)

	Mean % score in range				
Performance band	0–20	21–40	41–60	61–80	81–100
Most able (top 20%)	0	1	1	12	86
Above average	1	1	10	29	59
Average (middle 20%)	1	9	12	35	43
Below average	8	12	28	23	29
Least able (bottom 20%)	29	26	24	17	4

The differences between boys and girls in terms of overall mean scores for each form are not great (about two to three percentage points) and are not statistically significant. They are worth noting, though, because they repeat a picture already established for 11 and for 13 year olds, except in that the graphical advantage of boys appears to increase slightly between the ages of 13 and 15. This could simply be a consequence of the greater complexity in the questions administered to the 15 year olds – as noted earlier. It might also be a result of boys' greater take-up of Physics at 13+ (see Chapter 13 and Johnson and Murphy, 1986).

There are some clear and *general* differences between the performance levels of pupils in England, Wales and Northern Ireland. Specifically, English 15 year olds have as a group produced higher mean scores than their Northern Ireland peers on 65–75 per cent of all the questions, and have produced higher mean scores than their Welsh contemporaries on 75–85 per cent of all the questions. In both cases this pattern is already present to the same degree at age 13. At age 11 any differences between the average scores of pupils in the three countries were very much smaller, and were not always in favour of the English pupils. In particular, among 11 year olds Welsh pupils have shown the superior performances on questions featuring tables, while Northern Ireland pupils have shown a higher level of ability to handle coordinate graphs.

Why should 15 year olds in Wales, and to a lesser extent, in Northern Irish fail always to reach the levels of performance of their peers in England in this skill area? As with the gender differences in graphical skills, it might be speculated that differential national Physics take-up rates could hold a clue, since this subject more than any other offers extensive practice in this particular activity. Certainly, proportionally fewer pupils choose this subject in Wales and Northern Ireland compared with England once it becomes optional. But then the Welsh and Northern Irish pupils have demonstrated their relative weakness in this area at age 13 before option choices could have any influence. There must be some other reason, then, perhaps connected with differences in emphasis in the secondary school curriculum in the three countries.

The 1982 Mathematics survey throws some light on the problem. In this survey there was an enquiry into pupils' opportunity to learn various topic skills in

Mathematics. In general, for most of the topics included the English pupils were found to have the most experience, followed by pupils in Northern Ireland with the pupils in Wales proving relatively disadvantaged in this respect – particularly in the topic area **Probability and Statistics** which includes the 'representation and interpretation of statistical graphs'. One of the most striking national differences in topic exposure emerged for just this kind of graphical activity. Among the sample 15 year olds from England, Wales and Northern Ireland 8 per cent, 35 per cent and 6 per cent, respectively, claimed never to have met the topic. In contrast, 35 per cent, 17 per cent and 33 per cent, respectively, claimed to have worked on the topic during the Autumn term in which testing took place (Foxman *et al*, 1985, p574). Clearly, on this evidence, pupils in secondary schools in Wales engage very much less frequently in graphical activity than do their contemporaries in England and Northern Ireland, and this activity is of a less advanced kind. The relative weakness on the part of Welsh pupils in data representation skills is surely a direct outcome of this lower level of relevant experience throughout their secondary schooling.

Different curricular experiences in science *undoubtedly* explain the subgroup differences in performance which have emerged for the questions exploring pupils' familiarity with chemical symbols, circuit diagrams and section drawings. The least able pupils have fared particularly badly here, girls have shown less familiarity with the symbolic conventions than have boys, and pupils in Wales and Northern Ireand have been less successful than have their peers in England. Since the conventions concerned are taught only in the appropriate subject lessons in science, these differences in familiarity clearly reflect the differential take-up of the three main sciences by the various pupil groups (see Chapter 2).

5.9 Relative competence with different representational forms

Since the same data can often be presented in more than one way, it is tempting to ask to what extent pupils find the different representational forms themselves more or less difficult in comparison with each other. While tables perhaps provide the most versatile form in that many kinds of statistical and other information can be tabulated, charts and graphs have the advantage of immediate impact in their visual portrayal of any pattern in the data. Do pupils find graphs or charts easier to handle than tables because of this visual appeal? Unfortunately, it is not possible to answer this question from the evidence available in this science pool, for there are no questions in which the same data has been presented in different ways and the same tasks demanded of pupils. As mentioned earlier, the questions featuring tables, bar charts and pie charts have usually presented 'everyday' economic or social data while those featuring line graphs have usually presented physical science variable relationships. Moreover, the complexity of task demand has also been uncontrolled, so that while most of the bar chart and pie chart questions have been rather elementary in this respect, some of the line graph questions have posed inherently more difficult exercises.

Accepting this, there *is* evidence that while the more able pupils have coped equally well with all the different forms of data representation, the least able pupils have faced the greatest difficulties with the line graph questions – particularly those which required them to construct a graph given some tabulated data. However, bearing the previous caveat in mind, it does not follow that coordinate graphs are more difficult for these pupils to handle than are pie charts or bar charts. Were more complex pie chart or bar chart questions to be introduced into the pool, along with associated tasks of greater complexity, then the picture would certainly change. Indeed, an example of this is provided by the APU mathematics surveys. The evidence of these surveys was that pie charts proved the most difficult representational form for pupils (Foxman *et al*, 1985, p388). However, the few pie chart questions concerned seem to have been almost all concerned with relating sector sizes to quantitative amounts in a direct way, ie involving pupils in applying proportionality concepts. In these science surveys the pie chart questions have posed very much more elementary tasks and there is no evidence here that pie charts have proved at all troublesome – even to the least able pupils.

Thus, the *tasks* demanded of pupils in these questions are greatly influential, and little can sensibly be said about the relative ease with which pupils can handle different representational forms unless the same tasks are set with the same data presented in different ways. This issue was explored to a limited extent in the USA national surveys with rather inconclusive results (see NAEP, 1979a, p31). It seems that where elementary tasks are concerned, such as simply identifying the largest sector in a pie chart or the longest bar in a corresponding pictograph, then it matters little which form is used. However, when quantitative readings and arithmetic manipulations are required then both the representational form and the complexity of the data affect performance. Thus, in a pictograph it matters whether each symbol in a bar stands for one unit or for many, and in a bar chart it matters whether repeated charts are involved and whether bar lengths coincide with marked scale gradations. With the most complex data tables are often preferable to other forms; charts and graphs have their greatest impact where the data represented are straightforward and the variable relationships simple. So, without further investigation little can be said about the relative ease or difficulty with which 15 year olds might handle the different representational forms.

5.10 Summary

The majority of the 140+ questions on which this chapter has been based feature the most familiar and widely used forms of data representation, *viz* tables, bar charts, pie charts and line graphs. A handful of questions have explored pupils' abilities to handle less common forms of information representation such as Venn diagrams, flow charts and food chains, and the more specialist chemical symbols, circuit diagrams and section drawings. The information presented is most often statistical data of some kind, frequently economic or demographic in nature (typically claimed to be survey results) but sometimes scientific (experimental results, for instance). Frequency distributions have emerged as a popular type of data in the bar chart, histogram and to some extent the pie chart questions. On the other hand, the line graph questions are almost all concerned with temporal data, ie the graphs are typically temperature-time or distance-time graphs, or are cooling curves, solubility graphs, or growth curves. The majority of the line graph questions have featured the kind of data which would be most associated with the physical sciences – in contrast to the questions featuring other representational forms which have tended to present 'everyday' social or economic data.

The findings from these science surveys about pupils' abilities to handle the various representational forms are in line with those previously reported for the APU Mathematics surveys (see Foxman *et al*, 1985; and Cambridge Institute of Education, 1985), the corresponding national Mathematics surveys carried out in the USA (see NAEP, 1979a, b, c) and the CSMS study (see Kerslake, 1981). Together these various investigations have provided a rather detailed picture of pupils' capabilities in this area.

The evidence is that the majority of 15 year olds can successfully cope with the more general representational forms provided the associated tasks demanded of them are elementary. Thus, most 15 year olds – even the least able – show understanding that the lengths of bar lines, the size of pie chart sectors and the positions along a coordinate axis relate to quantity. They would, for instance, be able to identify 'the tallest pupil' in a bar line chart showing the heights of different individuals. They would be able to indicate 'the most frequent leaf width' given a histogram illustrating a frequency distribution of numbers of leaves within certain widths. They would be able to identify the food with the greatest proportion of protein, given a set of pie charts showing the proportional compositions of different foods. And they would be able to indicate the time of day at which the temperature was highest, given a temperature-time graph whose maximum coincided with a marked scale gradation on the time axis.

Most 15 year olds are also able to read or insert individual table entries as directed. They can also read bar lengths or point coordinates and can insert readings into bar charts or line graphs appropriately *provided* no interpolation of marked scale gradations is needed. Indeed, among the most able pupils (the top 40 per cent) few would find any difficulty at all with any representational task, however complex. It is the average and the least able pupils who experience problems as the tasks become less elementary. It is *these* pupils who are thrown when scale interpolation is required, or when the scale is more complex than direct counts in units or tens, or when two or more readings have to be arithmetically manipulated in some way (either simply added or subtracted), or when graphs are superimposed, or when pie chart sectors other than halves and quarters are involved, and so on. Whenever any one or more of these complications are present in a question performance levels have fallen among the weaker pupils. It is also the less able pupils who have faltered – to the extent often of not responding at all – when faced with questions involving the very science specific symbolism. Not surprising this, since, as Chapter 2 has shown, few of these pupils study Chemistry or Physics.

As far as gender differences are concerned, the data suggest that 15 year old boys are slightly more able than girls of this age to handle line graphs successfully, while girls have retained an earlier slight advantage when it comes to tables and charts. Also, English pupils have performed uniformly more successfully in general at this age than have their peers in Northern Ireland or, particularly, Wales. These are likely to be outcomes reflecting curricular differences between these groups of pupils. This is particularly so as regards Physics take-up, since this subject more than any other relies heavily on data representation skills and therefore provides continuous reinforcing practice both in reading *and* in constructing graphs and charts (for further discussion see Chapter 13).

6

Use of apparatus and measuring instruments

6.1 Introduction

This chapter reviews the performance of pupils across the range of abilities described by **Use of apparatus and measuring instruments** and includes a brief reśumé of the tests used and their administration.

For the first survey in 1980 the following distinct aspects of relevant practical activity were identified (Driver *et al*, 1982):

Using laboratory instruments – pupils required to read instrument scales, to apply appropriate units and to use measuring and laboratory instruments correctly.

Estimating – pupils required to estimate the magnitude of physical quantities by judging their value, or by attempting to take a specified amount of material.

Following instructions – pupils required to follow written instructions and/or diagrams. Following the instructions may require simple observation, recall of procedures or exact compliance with a set of directions.

(The process of evolution of the list above is described in the age 13 review report—Schofield *et al*, 1988—and is not repeated here.)

These components of the category have been included in each of the surveys of 1980, 1981, 1982 and 1984. (As part of the overall strategy of the monitoring programme **Use of apparatus and measuring instruments** was not included in the 1983 survey.)

Cost and logistic constraints preclude the administration of large numbers of questions from this category in any one survey. In the first annual survey 29 questions were used and in the second 18 questions were administered. Such constraints also impose some restrictions on the range of apparatus and measuring instruments which could be used in testing – for instance, the precision 'top pan' balance which is too costly and difficult to transport has not been included in testing.

Moreover, there is only a limited number of instruments and apparatus for which it might be considered educationally sensible or useful to assess pupil competence. The range and number of genuinely different questions which can be envisaged to test the activities listed above is also limited. It is, for instance, by no means clear that a question involving the use of, say, a measuring cylinder to measure a volume of one fluid is a genuinely different question from requiring the measurement of a volume of a different fluid.

Questions from the three aspects of this category were therefore incorporated into a fixed test described in detail in the report of the 1982 survey (Gott *et al*, 1985). That report also outlined the collection of information on pupils' responses and described the way in which the test score was produced.

In 1984 the fixed test was modified to permit comparison with the performance of 13 year old pupils on the same test. Adjustments were made to questions used previously at one age or the other, and the time spent by a 15 year old pupil at each station of the circus was increased from six minutes to eight minutes. Discussion of the comparative performances of 15 year olds and 13 year olds is presented in Chapter 12. Appendix 5 of this report carries the 1984 performance data for 15 year olds; that for 13 year olds is presented in Chapter 6 of the age 13 review report (Schofield *et al*, 1988).

6.2 Test administration

In four annual surveys more than 8000 pupils from 484 schools in England, Wales and Northern Ireland have attempted circuses of questions testing **Use of apparatus and measuring instruments.** The approach adopted in the surveys has been to assess pupils in groups of nine each working individually around a circus of experiments arranged in nine or ten stations, moving from one station to another in a set sequence at timed intervals. The tests have been set up and run by trained testers/supervisors. These supervisors have taken to each school all the equipment needed and administered the tests according to instructions received during training. They were helped in servicing the circuses and in gathering the more ephemeral pupil performance data on checklists by one of the school's own science staff released by the school for the day's testing. Full details of the recruitment and training of senior science teachers as supervisors, and of the provision of equipment by Philip Harris Biological Ltd, are given in previous reports (eg 1980 survey, Driver *et al*, 1982) and in the Science Report for Teachers: 6, Welford *et al*, 1985.

The list of apparatus and instruments used in the surveys was presented in Appendix 1 of the 1980 report (Driver *et al*, 1982, pp203–205). This list was compiled with the assistance of more than 150 science teachers with different discipline backgrounds from different schools and regions of the country. They also suggested laboratory procedures that 15 year old pupils would be expected to know by the time they leave school.

Consistency and reliability of the tests

Much thought and effort has been expended in establishing test conditions which are as near as possible identical for all pupils taking part in the circus practicals. Debate prior to the first survey in 1980 considered the alternatives of calling upon survey schools to provide the equipment and apparatus for testing or of centralising and standardising the provision of equipment. The former has the advantage of allowing pupils to operate with familiar equipment; meter faces for instance show a wide variation in scale and design. However, the decision was to adopt the latter approach, and the provision of standardised test kits put together by Philip Harris Biological Ltd to the monitoring teams' specifications has eliminated the variability (of unknown magnitude) which would have resulted from non-uniform equipment. In addition, the not inconsiderable additional burden which the former approach would have imposed on schools has been avoided.

The recruitment and training of experienced science teachers to administer the tests has been aimed at reducing the effect of variation in administration. At training all aspects of the questions, organisation of pupils, collection of pupil responses and of actual set values were rehearsed. Supervisors practised maintaining values of, say, temperature in a flask over the test period, and of recording the values (if any variation occurred) encountered by every pupil in the test. In this way it has been possible to predict and adjust values *in situ* and to chart the possible sources of error. As a result pupil scores have been calculated on the basis of the deviation of their response (recorded value) from that which they encountered and not that which they were assumed to be meeting.

This is not to claim that every circus in every school has run without interruption or mishap, but since any hiatus, value slippage, equipment failure or the very occasional instances of problems associated with pupils will have been noted, it has been possible to adjust or even discard suspect data.

It has also been straightforward to establish that training instructions have been adhered to and that consistency of testing has been achieved between schools, regions and countries included in the surveys. The lack of variability in results to the same question set in different years to different yet representative samples of pupils attests to the effectiveness of the test administration.

Supervisor values collected do show that in a small number of cases, even with groups of nine pupils in tightly controlled test circuses with a member of the test school science staff in attendance, the set value demanded for a question has not been achieved. This would suggest – since such a deviation is usually annotated by the supervisor affected – that test circus administration is stressful and difficult. A pupil accidentally upending a set of apparatus at one station (a very rare event) will cause an upset to routine which has resulted in the supervisor not having time to adjust some value at another station. While such events have been recorded (and resulting suspect data eliminated) in APU testing of *populations*, the implications for summative (or formative) in-school assessment of *individuals* are clear.

Two potential sources of variation in pupil performance arise from internal characteristics of the questions themselves or from the order in which pupils proceed around a circus of questions. In 1984, for the first time, the internal order of multipart questions was varied to minimise the effect on scores of pupils running out of time. Analysis of question mean scores and of test mean scores has shown that there is no difference in performance between pupils meeting questions in alternative internal orders.

The fixed test in this category includes a variety of questions in which pupils are required to use or read instruments at some circus stations, and at other stations to make estimates of physical quantities without recourse to instruments. Since candidates were moved around the circus in a set sequence some will have met 'estimation' questions before they had the chance to 'use instruments', and some will have been required to 'estimate' after 'using instruments'. Again analysis of pupils' performance has shown no significant difference attributable to circus order in overall test scores or individual question mean scores.

6.3 Pupil performance on 'Use of apparatus and measuring instruments'

This section reviews pupil performance across the variety of tasks which make up the fixed test of **Use of apparatus and measuring instruments** and by drawing on data from all surveys to date presents a description of common errors and other features which characterise responses.

Reading scales on pre-set measuring instruments

Arguably the most straightforward of tasks within this category is that of simply reading the scale of a measuring instrument pre-set at a fixed value. Pupil performance on this activity was presented in the reports of the 1980 and 1981 surveys, and again in considerable detail following the 1982 survey. The more detailed marking reported for the 1982 survey was repeated in 1984 and the 1984 results are shown in Table A5.1 in Appendix 5.

Some of the instruments were set at slightly different values in 1984 from those adopted for the 1982 survey to ascertain any effect of change in set value on performance. The scale reading question used in 1984 was very similar to that described in the 1982 report (Gott *et al*, 1985, p27). The sources of error described in previous reports were again detected in pupil responses to this question, and these are now summarised for each instrument. The reader should refer to Chapter 3 of the 1982 survey report where most of the instruments are drawn to show the scale and scale intervals in use in the fixed test.

Reading a measuring cylinder

Performance has varied with the set value used. However, regardless of that value, of the order of twice as many pupils read the *top* of the meniscus as read the set value. A significant proportion of pupils appeared to round up or down to the closest major scale point. The actual set value influenced the magnitude of this error. A value of 53 cm^3 had 13 per cent of pupils rounding up to 55 cm^3; a value of 42 cm^3 resulted in fewer than five per cent of pupils rounding up to 45 cm^3. Parallax errors were detectable in 10 per cent of the responses in 1984.

Many pupils appeared to mistake the position of the major scale indicator and gave a value which suggested that they were reading the scale in the wrong direction.

Reading a forcemeter

As with the measuring cylinder, performance varied with the set value. If set on a major and labelled scale line – 30 N in 1980 and 1981 – then the value was read accurately by well over 80 per cent of 15 year olds. If on a scale line (unlabelled) – 16 N in 1984 – then around 80 per cent of pupils were within one scale division of the value. If between scale lines (eg 17 N in 1982), so requiring interpolation, then fewer than 70 per cent of pupils were within the same range of tolerance. The only consistent and major source of error was that of counting the minor scale interval (2 N) in ones not twos. This error was common for all instruments where the minor interval was two units.

Reading a manometer

In the fixed test of 1982 the manometer was the last of 10 instruments to be read, and since 43 per cent of pupils failed to record an answer it was assumed that some pupils ran out of time before attempting to record a reading for the manometer. To counter this the order in which the pupils met the nine instruments to be read in 1984 was varied so that no one instrument was always read last by all the pupils.

However, the water manometer was not read accurately, regardless of its set value, with only between 10 per cent (1982) and 22 per cent (1984) of pupils doing so to within 1 cm H_2O of the set value.

It was suggested in earlier reports that the need to compute a difference between two readings rather than to record a single value would cause difficulty. In 1984 pupil responses were analysed to shed light on the extent to which this requirement was a major problem to pupils. Just under two per cent of pupils recorded the heights in *both* arms of the instrument, 24 per cent gave the *upper* reading only, and six per cent gave the *lower* reading only. While 45 per cent gave a reading which was clearly an attempt at computing the difference between the two values, the value given by half of these pupils was not categorisable and could have been either a gross misreading or a very inaccurate arithmetic substraction involving two whole numbers.

Reading a thermometer

Clearly and consistently, regardless of set value, reading this instrument has provided fewest problems, with between 82 per cent and 86 per cent of pupils able to record a value to within 1°C of that set. Errors fall into no consistent pattern other than a small proportion of pupils who rounded up or down to the nearest major marked scale point.

Reading a stopclock

Faced with a clock stopped at 7 minutes 17 seconds many pupils recorded 6, 7 or 8, 17 or 7.17, accompanied by the units expressed either in minutes or seconds. Pupils have often mistaken the position of the minute hand or have been unclear as to which of the two hands was the minute hand and which was the second hand; an error which might not have arisen were the pupils able to see the hands moving. However, about 55 per cent of 15 year olds could read a pre-set stopclock to within one second of the set value.

Reading an ammeter

The two most common sources of error were associated with characteristics of the scale of the ammeter used (adapted to show only a single scale with clearly marked major scale intervals). Pupils found minor scale intervals advancing in twos difficult, and counted in ones. The added complication of a second place of decimals (eg set value 0.24 A, minor scale interval 0.02 A) produced many answers ten times the set value – most commonly recorded as 2.2 A (counting in ones) or 2.4 A. Fewer than one in five pupils read 0.24 A to within 0.01 A.

Reading a voltmeter

Although essentially the same scale face as the ammeter the minor scale interval was 0.1 V and so neither of the problems of counting in twos nor manipulating the second place of decimals occurred. Consequently performance was much better, with between 60 per cent (1982) and 84 per cent (1984) of pupils within 0.1 V of the set value. Choice of set value can, however, affect the facility with which this instrument is read. A value of 1.1 V (1984) or 3.4 V (1980 and 1981) meant that rounding down or up to a major scale point was within the range of tolerance used (the minor scale interval of 0.1 V). 1.3 V ± 0.1 V (1982) did not embrace those readings of pupils whose perceptions of the accuracy necessary to the task permitted them to use the nearest major scale point as sufficiently accurate, and showed a reduced proportion of the sample within 0.1 V of the value set.

Reading a lever arm balance

It is possible that unfamiliarity with this instrument may account for the wide distribution of pupil values around the set value with about 40 per cent of 15 year olds reading the value accurately. Similar errors have been noted in each of the surveys (for example, in 1984 the set value was 117 g):

- a tendency to round up or down (to 120 g in 1984);
- parallax error in reading (114, 116, 118, 119 g in 1984);
- misread or misrecorded (as 107 or 108 g in 1984);
- counted in 'ones' not 'twos' (minor scale interval – 2 g);
- recorded convenient whole numbers or marked major scale points (100, 110 g in 1984).

Reading a rule

Perhaps a surprising result in view of the familiarity of the ruler to pupils is that reading a set value has produced a relatively low performance figure (c 50 per cent). In 1982 arithmetic manipulation (a subtraction was involved) was suggested as reducing the percentage of pupils who used the ruler accurately. In 1984 the need to subtract was eliminated, but the value to be read was 30.3 cm. Again it may be that pupils' perceptions varied as to the accuracy demanded – many rounded down to 30 cm. Manipulation of decimals (although only one place) has proved an obstacle with 20 per cent of pupils recording 33.0 cm! This is not an unexpected mistake but noteworthy for the large proportion of pupils making such an error.

Paper and pencil testing of reading scales

A comparison of pupil performance on reading instruments set at certain values with that on reading photographs of instruments showing the same set values was described in the report of the 1983 survey (Welford *et al*, 1986). Similar work has been done to look at performance on line drawings of the same set values. There is a very close match between practical performance and that on the photographic version although parallax or perspective problems interfere with the clarity of the photography. Line drawings may overcome such hurdles, but while performance is slightly better overall, it is very sensitive to the *quality* of the line drawings of most instrument scales. Paper and pencil versions are no substitute for practical experience in *learning* how to read instruments, but quality photographs seem to be cheap and accurate substitutes for testing the acquisition of scale reading skills. Line drawings provide an inconsistent though usually small overestimate of performance. Reading scales of pre-set instruments is, of course, a task not set in the context of activity, and with little in the way of feedback which might result from *using* the instruments in tasks (however limited) or in more holistic investigative practical activity.

Using instruments to measure quantities

Several questions have been used to test this ability, but there are two main demands framed by the tasks:

- Can pupils use various instruments to measure known values, such as the extension of a spring, or the period of illumination of a flashing light?
- Can pupils use instruments to deliver a certain quantity of a substance, for example, weigh out a quantity of sand, or cut a prescribed length of paper tape?

Such tasks have an explicit though limited purpose, but one not subsumed within a larger more investigative operation, such as the 'complete investigations' which are described in Chapter 11 of this report.

Results revealed below show remarkable consistency from year to year, with performance variation dependent upon the physical quantities to be measured and between the task demands outlined above.

Measuring length

Measuring the extension of a spring involves two readings and the subtraction of one value from another. The proportion of pupils giving an accurate reading (within 1 mm on a 40 mm extension) was low (35–40 per cent) compared with the proportion who could accurately cut 473 mm of paper tape to within 1 mm (60–65 per cent). Errors in the first task have been:

- incorrect subtraction of the initial from the final length (160 mm from 200 mm);
- recording either the initial or the final length of spring without subtraction.

In the second (paper tape) task the errors would seem to have been mostly associated with manipulation, and cutting obliquely across the tape. Most of the lengths delivered were only a few millimetres outside the limits set of ± 1 mm. In 1984 for instance 63 per cent of pupils were very accurate (472 mm–474 mm) and nearly 90 per cent cut lengths between 468 mm and 478 mm – ie within 5 mm.

Measuring volume

The task which demanded that pupils measure the volume of water needed to fill an ungraduated plastic cup up to a fixed mark (136 cm^3 in 1984) was completed with much less accuracy than that in which pupils were required to leave a measured quantity of water in a measuring cylinder. In the first task only 14 per cent of pupils were within 1 cm^3 of the value, whereas fully 61 per cent of the same pupils could deliver 55 cm^3 of water within the same limits of accuracy. In both tasks a consistently large proportion of pupils used the top of the meniscus as the level to read, but whereas this was the only large source of error in the latter activity, in the former task could it be that pupils' perceptions of the accuracy needed have affected the reading recorded? It was certainly the case that the strategies adopted to complete the exercise meant that over half of the 15 year olds tested used convenient major scale indicators such as 120 cm^3, 130 cm^3 or 140 cm^3.

Measuring mass

Consistently around 40 per cent of pupils read the lever arm balance to within one minor scale gradation of a set value. However, involvement in a more active task results in performance variation which has seemed to be more dependent upon the material being weighed (and the consequent and different manipulations of both apparatus and arithmetic) than on the balance or physical quantity itself. Plasticine, sand and water have all been used for this activity. Results suggest that about two-thirds of pupils can weigh out a given mass of plasticine within the acceptable limits of accuracy; one-third can weigh out a set mass of sand and just over one-fifth of pupils can weigh out accurately a given mass of water.

Measuring other quantities

The range of instruments for which it is feasible to make comparisons of tasks such as those described above is limited especially within the constraints of large scale survey testing. However, over 70 per cent of pupils determined accurately the force required (5.5 N) to lift a small mass and time accurately an event of fixed duration (4.0 s). When required to measure a certain volume (1.4 cm^3) using a syringe, fewer than a fifth of pupils positioned the plunger correctly, and nearly a half had little or no obvious idea of how to use a syringe to deliver an accurate volume of a liquid.

Using a hand lens and a microscope

Pupils were given three slides on each of which was mounted a different set of four letters of the alphabet, much reduced. They were asked to view the first slide using a hand lens, focus a microscope on the second slide already placed and centred on the stage, and then to stage, centre, focus and view the third slide. In each case the pupils were required to copy what they saw (they were not told that they were looking at letters). Perhaps surprisingly 15 year olds have been consistently better at using the microscope than the hand lens. Around 80 per cent of pupils were completely successful when carrying out the two tasks with the microscope, with no outstanding source of error among those unable to complete the tasks. By way of contrast, fewer than two-thirds of these pupils showed such proficiency with a hand lens and commonly misplaced letters or mis-orientated the whole sequence of characters.

Using apparatus in standard techniques

Pupil performance on a variety of different tasks using repeated measurements and involving the correct use of standard apparatus has been well described in previous reports and in the Science Reports for Teachers: 5 and 6 (Gamble *et al*, 1985; Welford *et al*, 1985). For the most part the general level of performance, judged both by the final accuracy of the measurement, and the technique of operation, is not as good as might have been expected. Paper and pencil (photographic) versions of such questions suggest that the actual practice of *performing* these tasks would seem to be less well known than the *theory* of how to carry them out (Welford *et al*, 1986).

In addition to the problems just described, pupils have displayed a number of errors in using a lever arm balance (Driver *et al*, 1982). They have failed to centre the weighing pan, failed to empty the pan prior to the next weighing, failed to ensure that the apparatus was working without sticking and was zeroed at the start.

A question which required pupils to measure a small time interval – the period of a pendulum (Driver *et al*, 1982, Example 4, p50) has been put to different samples of 15 year olds in different surveys in slightly different guises, but always with the same result. Between one-half and two-thirds of pupils attempted to time one swing only. Whereas most of the rest used methods based on timing a number of swings, fewer than one-third of pupils ever attempted to confirm their readings by repeating or improving their chosen method. Nearly one-fifth of the pupil sample failed to stop their stop-clock before attempting to read it. A similar proportion swung the pendulum in such a way that it was impeded by the suspending apparatus. On a more positive note, very few pupils actively pushed the pendulum bob, or started their stopclock arbitrarily. Those who repeated the operation tended to improve their technique in subsequent attempts.

In the same vein, most pupils used a single bung to determine the average mass of each of 100 such bungs (where a single bung was too light to effect a single minor scale deflection), a single bean to establish average length given many beans of uneven length, or a single marble when given several and requested to find the average volume of the marbles.

Other features of note are:

- most pupils (around 90 per cent) ensured that a measuring cylinder was vertical before reading it;
- between 70 per cent and 80 per cent of 15 year olds read a measuring cylinder with their eye at water level;
- a quarter of pupils failed to take a thermometer out of its plastic case when measuring the temperature of water (though only one per cent read it upside down!);
- nearly a fifth of pupils withdrew the thermometer bulb from the water before reading;
- one in ten pupils failed to wait for the thermometer to reach a stable final temperature, but read it immediately upon immersion.

Estimating physical quantities

In the laboratory or science lesson pupils are frequently required (usually informally) to 'guesstimate' the magnitude of physical quantities, or to be aware of the relative amounts of various quantities being described or used. Such a 'feel' for quantities is essential for selection of appropriate measuring instruments, for attaching appropriate units to quantities and so on. A range of questions of both the broad types described in the introduction to this chapter have been used, namely those requiring pupils to estimate the magnitude of physical quantities and those which ask for pupils to take a specified amount of a material. The results have shed light on just what sort of meaning 15 year old pupils attach to the magnitude of physical quantities. The 1980 survey report (Driver *et al*, 1982) presented a detailed breakdown of pupil performance on estimating physical quantities. The results from both the 1982 and 1984 surveys are presented in Appendix 5 of this report, showing the proportions of pupils who made their estimates to within 10 per cent, 20 per cent and 50 per cent of the given or requested value. These ranges are chosen to allow the reader to judge pupil competence against more than one yardstick of accuracy and allow a variety of interpretations or perceptions of 'good' or 'poor' estimating.

It is worthy of note that the more active and participatory task of attempting to take a specified amount of a given material generally resulted in closer estimates than 'guesstimating' given quantities. In the former task (although not all the physical quantities tested in the latter are included), between half and three-quarters of pupils presented a quantity which was within 50 per cent of that requested. In contrast, only length, time and temperature were estimated as proficiently in the second, passive, task.

The pattern which emerges from the data for the 'guesstimate' type of task is that three-quarters of pupils made estimates within 50 per cent of the set value for length (straight or circular), half were similarly accurate for time, temperature, or small areas, about one-third were as good at estimating the quantities of volume, larger areas, small masses and force, and pupils were least successful at estimating the value when given a large mass to handle.

The tendency to over- or underestimate would appear to be at least as dependent upon the value being estimated as upon the physical quantity under consideration. Very small amounts were consistently overestimated but larger amounts were usually underestimated.

Use of units

The computation of question scores in this category has always included a component awarded for the attachment of suitable units to all values recorded. Questions involving estimation have given pupils the units relevant to the physical quantity under consideration. Scale reading questions have reminded pupils to say which unit they were using but other measurement questions have not reminded or directed pupils to accompany their answers with the appropriate units. In 1980 some questions were administered which specifically tested pupils' knowledge of units (Driver *et al*, 1982, p46). In marking pupils' written answers to the practical questions set in both the 1982 and 1984 surveys, pupils' use of units was specifically coded as described for the 1982 survey (Gott *et al*, 1985, p38).

It has been noted that, in general, those questions which asked pupils to choose the unit to go with the instrument had higher mean scores than those which asked for the instrument that went with the unit (Driver *et al*, 1982, p46). The main findings reported for the 1982 survey (Gott *et al*, 1985 pp38–39) were confirmed by the 1984 data. The reminder to use units always resulted in increased use of units. Irrespective of reminder or direction (or the absence of such) the proportion of pupils who used incorrect or unsuitable units, or who used units interchangeably, was very low. Pupils more usually did not accompany values with units rather than use units incorrectly. The frequency of use of the SI notation for symbols would seem to be dependent upon the physical quantity involved, and to some extent upon the pupils' familiarity with the instrument being used to measure a certain physical quantity (volume measured using a syringe rather than a measuring cylinder was characterised by a decreased use of *any* units, and a decreased use of the SI notation). For mass, length, temperature and volume involving the use of a measuring cylinder 80–95 per cent of the units which were recorded employed the SI notation. SI units and symbols were not nearly as widely employed for the other physical quantities tested, with between 15 per cent and 45 per cent of pupils recording these units demonstrating knowledge of the SI conventions.

Following instructions

Every question used in APU practical science has tested pupils' ability to follow instructions. As described in the introduction to this chapter, pupils have been specifically tested on whether they can follow written and/or diagrammatic instructions. Performance on following written instructions was reviewed in detail for the 1980 survey (Driver *et al*, 1982, pp59–64), and on following practical instructions using conventional symbols for the 1982 survey (Gott *et al*, 1985, pp40–43).

In order to evaluate pupil competence in proceeding through a series of instructions it has usually been necessary for the supervisor or the participating school's teacher to complete a checklist for each pupil. Such a constraint has limited the number of questions set to test this aspect of practical work, but it has also been possible to collect behavioural information about pupil activity which has enabled judgements to be made as to pupils' effectiveness in carrying out practical tasks.

While it is acknowledged that test conditions have meant that perhaps pupils take more care to read and to proceed logically through a set of instructions, and that they are less likely to treat the questions as 'just another worksheet', between 80 per cent and 90 per cent of the 15 year olds tested have demonstrated the ability to work systematically through a given sequence of directions. The cause of pupils failing to complete a stage in an experiment has more to do with a lack of manipulative skills, irrelevant observations, or failure to recall appropriate techniques than with an inability to follow instructions. For instance over 80 per cent of pupils were able to comply with the request to cut off a small piece of material, immerse it in a drop of water on a microscope slide, cover it with a coverslip, place the slide on a microscope stage and focus at low power. The accuracy of cutting the material, the exclusion of air bubbles and the presence of water in places other than under the coverslip may leave much to be desired, but such faults are not to do with failing to follow instructions.

There is evidence of relatively poor performance on following instructions based on the application or use of taught science knowledge. For example the instruction to set up an electrical circuit to match a circuit diagram was carried out successfully by only 14 per cent of all pupils (27 per cent of those studying Physics). Replacing the circuit diagram with a set of written instructions resulted in a similar lack of success. However when pupils were provided with a photograph of the circuit about 70 per cent correctly produced every detail of the required circuit (Gott, 1984).

6.4 Pupils' test performances

The purpose of this section is to describe characteristics of pupils' performance on the fixed test as a whole. The 1984 fixed test results are used throughout this section as representative of performance patterns in this category. (The more detailed discussion of performance on components of the test has been presented above.)

As has been reported for all the science surveys to date participating schools complete a questionnaire providing background information relating to the school and to each pupil selected for testing.

Pupil performance on the whole fixed test according to the variables of gender, science curriculum followed in their fifth year, and school estimate of projected examination entries is presented in Table 6.1 below.

Table 6.1 *Pupil performance levels on the whole fixed test broken down by examination entry, gender and science curriculum*

(Unweighted mean percentage scores, 1984 survey)

	Mean % score		
All	All pupils (n=612)	Boys (n=314)	Girls (n=298)
8+ O-levels	61	63	60
6–7 O-levles	57	60	54
3–5 O-levels	54	57	52
1–2 O-levels	52	52	52
4+ CSEs/no O-levels	46	49	42
0–3 CSEs/no O-levels	39	39	38
Studying Biology	56	58	55
Not studying Biology	49	51	46
Studying Chemistry	60	60	61
Not studying Chemistry	49	50	48
Studying Physics	59	58	61
Not studying Physics	48	48	48
All pupils	52	53	51

The better performance of boys in this area of scientific activity has been noted in previous APU survey reports, and is again evident here. The NAEP (National Assessment of Educational Progress) surveys have also found boys to be significantly more competent at all ages in the use of apparatus and measuring instruments (NAEP 1978a, 1978b). Table 6.1 shows that the difference in favour of boys, while not large, is consistent across the exam entry groups, and for pupils studying Biology. However, girls studying Physics out-performed their male counterparts.

Performance comparisons involving the curriculum background of pupils, their gender or their ability are not straightforward. Looking now at the fixed test performances of the curriculum groups shown in Table 6.1, it is worth remembering that both the gender and ability of pupils are in interaction with their curriculum when comparing scores. It has been pointed out in previous reports (eg Driver *et al*, 1982) and in Chapter 2 of this report, that the girls who study Physics are, as a group, very able and that a majority of pupils who study Biology are girls who are fairly evenly distributed across the ability groups. Furthermore, the majority of these girls take Biology as their only science. Such a perspective could be used to explain the lower overall performance of those who are studying Biology. It may also be argued that a test of **Use of apparatus and measuring instruments** has a physical science 'feel' to it and that the biologists (many of whom study only this one science) could therefore be disadvantaged through a lack of experience in measurement. They certainly performed less well across the fixed test than did their peers studying the other sciences.

Within each range of ability (judged by potential examination entry) boys outscored girls. In fact, from Table 6.1 it is evident that the boys from any examination entry group were at least the equal of the girls from the immediately higher group. This may reflect the interactions between gender, ability and curriculum patterns referred to in the previous paragraph. It must also be explained, in part at least, by the lack of weight given in more conventional examinations to the abilities tested in this category. Such abilities may not therefore figure largely in schools' estimates of pupils' potential examination success.

An exploration of the match between examination entry and performance on this category is presented below in Table 6.2. Pupils have been grouped into five performance groups of approximately equal size based on their percentage mean scores on **Use of apparatus and measuring instruments**, and the proportion of each of these five groups included within each examination entry group is shown in Table 6.2

Table 6.2 *The distribution of pupils within each test performance group among the examination entry groups (1984 survey)*

	% pupils (n = 612)				
	Test performance group (test score range*)				
	least able (7–40)	below average (40–49)	average (50–57)	above average (57–64)	most able (64–88)
8+ O-levels	6	13	20	37	56
6–7 O-levels	8	6	18	10	17
3–5 O-levels	6	16	21	20	11
1–2 O-levels	10	8	11	13	6
4+ CSEs/no O-levels	38	45	22	17	9
0–3 CSEs/no O-levels	31	11	7	4	0

*Rounded figures.

The pattern of distribution of pupils shown in Table 6.2 is fairly predictable, but perhaps not as tightly grouped as might have been expected. Over half of the top performance group (56 per cent) fall into the '8+ O-level' ability class, and the least good test performers are also estimated to be least able by the schools.

However, the fit is not exact and there are a significant number of our school pupils, thought of as the most able, not performing well in this science activity, as well as many who might be judged as of lesser calibre who outscore others thought of as more able.

However, in the light of the earlier comments relating to the interactions between gender and curriculum and ability it may be of some interest to look at the distribution of boys and girls of various curriculum backgrounds within the five performance groups.

Chapter 2 has shown that many of the pupils of 8+ O-level estimated potential study more than one science, while the majority of those of lesser ability study only one of these three sciences, or General Science, while a minority are taking no science at all. The majority of the pupils studying any one (or combination) of the three named science subjects are located in average or above average performance groups. The unevenness of the distribution is most marked for girls studying Physics or Chemistry. It has been remarked earlier in this section that the girls who do Physics are the most academically able; certainly the girls doing physical science are high achievers on this test – 73 per cent and 71 per cent of girl physicists and girl chemists respectively are in the top two performance groups, compared with 51 per cent of girl biologists.

The group sizes (by gender, ability and performance) are too small for detailed comparison but almost all the girls studying physical science are from the 8+ O-level examination entry group and the top two performance groups. A not insignificant number of the girl biologists who are of 8+ O-level potential are in the bottom two performance groups. It is largely these pupils who make up the top left hand corner of Table 6.2!

6.5 Summary

The lasting impression of pupil performance in this category is of overall positive achievement – a category mean score of over 50 per cent – but an underlying disquiet which suggests room for much improvement in particular areas of the activities described. The final statement to the equivalent chapter in the 1980 report (Driver *et al*, 1982) was that '.....the knowledge of some of the basic practical skills is not as widespread as might have been anticipated'. Subsequent surveys have served to describe the peaks and troughs of pupil performance, and have attempted to ascribe causes to the effects noted.

It is quite apparent that scale related idiosyncracies of most of the measuring instruments used in the tests (and in everyday use in school) disturb performance significantly. Minor scale divisions mounting in 'twos' and manipulation of decimals (especially to two places) are massive hurdles for pupils trying to make accurate readings of instruments.

Results show that the actual location of the value to be read – pre-set or obtained during an experiment – disturbs performance. A value situated on a linear scale major marked division causes little perturbation; but the reality of experimentation is rarely so accommodating. The decline in performance deriving from more inconveniently situated values is marked.

Paper and pencil alternatives (photographs or line drawings) to practical questions have been shown to produce quite different pupil performance patterns when compared to the practical versions, except in the case of reading high quality photographic representatations of most instrument scales set at specific values.

Pupils recorded responses which can be thought to demonstrate a variety of perceptions of the accuracy needed to satisfy the demands of a question. The tendency to round up or down to the nearest whole number or marked scale division has a significance which will escape few teachers of science. Clearly although the questions are quite explicit in their demands pupils are either careless in their interpretation of accuracy required, or they are not familiar enough with these types of tasks to know when to be accurate.

In Chapter 11 it is argued that it is more reliable to assess pupils' abilities to use apparatus and measuring instruments by way of the 'atomistic' tasks which are employed in this category. Largely, this is because in the longer 'holistic' investigations pupils' perceptions of accuracy needed might vary, and probably vary with the type of approach they have chosen to use in solving their problem. Perception of which apparatus to select for use also varies, so making it very difficult to judge all pupils on their use of any particular instrument since not all pupils make the same choices.

However, even within the relatively limited scope of the tests of **Use of apparatus and measuring instruments**, tasks vary, performance varies with task, and with the degree of active involvement required of the pupil in the task. For instance, results reviewed in sections 6.3 and 6.4 established that using an instrument to deliver, say, a stated quantity of water had different performance characteristics to those of measuring the capacity of an ungraduated beaker.

The tension which exists between maximising the reliability of a test and maximising its validity has been well rehearsed elsewhere (see, for example, Woolnough and Allsop, 1985; Fairbrother, 1978). The tests used in this category would seem to satisfy most criteria of reliability. Results achieved in different years are largely comparable. Administration of the tests, given the intensive training of supervisors has been shown to be reliable. However, comparisons between the different types of tasks show there to be a certain task dependence which would not allow performance on a particular skill tested under one circumstance to be compared with that obtained on a slightly different task. The range of tasks across which the questions have ranged in this category, however, provides the test with a greater validity than would be possible from any more narrowly defined test. The validity of the activity upon which the test is based can be further increased by moving to assessment within whole investigations, with the consequent loss of reliability pinpointed in Chapter 11.

The age 11 review report (Russell *et al*, 1988) suggests that the performance of younger pupils is closely tied to the specific content of testing in this category. The report concludes that a more general use of instruments over a wider range of learning situations is necessary, and that 11 year old pupils will only gain an increased understanding of the appropriate use of apparatus and instrumentation through broadening the learning experience. The wider experience aimed at may give pupils insights into applications of measurement strategies of which the majority of these younger pupils seem at present to be unaware. Older pupils will certainly have studied more science and used a greater range of apparatus in a variety of situations. However the evidence advanced at age 15 may serve the same argument and call for the learning experiences of applications of measurement strategies to be as broad as possible.

The temptation is to conclude that the abilities being tested in this category are not specifically taught or systematically practised in years 4 and 5 of many secondary schools. Indeed, weight is lent to this notion by HMI (DES, 1979b) who noted that the proportion of science lesson time spent on practical activities decreases as pupils progress through the secondary school. Recent initiatives are perhaps seeking to reverse this inclination. A common theme running through 'Science 5–16' (DES, 1985), the National Criteria for Science (JCNC, 1985) and the draft reports of the Secondary Examinations Council Working Parties on grade related criteria in Physics, Chemistry and Biology is the renewed emphasis on 'doing' science.

It may well be argued that the content-laden state of science subject syllabuses leaves little time for the development of the abilities described and tested by the APU. It is probably true that such abilities are not tested (directly or indirectly) in the majority of science examinations at 16+, and that their teaching is of a lower priority than that aimed at the mastering of the materials which are examined. The rather inexact match between performance on the fixed test in this category and the ability of pupils based on their potential examination entry would support this view. The directive from the National Criteria for Science that 20 per cent of marks be awarded on the basis of practical work does now place a value on the acquisition of the abilities needed for successful practical accomplishment in science. It is to be hoped that the requirement to assess practical work will produce coherent teaching, learning and assessment strategies incorporating a systematised and rewarded development of practical abilities.

While the implications for learning are necessarily speculative those for assessment are less so. Summative internal assessment of practical abilities will attempt to be both reliable and valid and should therefore be based on a multiplicity of tasks. Pupils will need to be judged on the basis of a variety of 'atomistic' tasks, during more 'holistic' investigations and during other practical work. It is not hard to imagine the tests described in this chapter transferring quite easily to in-school testing. But, in the context of school testing, how educationally valid are such tests? The obvious danger is that the adoption of these relatively reliable tests will result in in-school assessment coming to comprise mainly (or totally) questions probing these 'atomistic' or 'particulate' skills. In addition, and of greater concern, since assessment can shape the curriculum, taught science could soon have a threadbare look and become a series of arid and featureless measurement activities. The need for the variety in testing argued above takes on added significance. Indeed the conclusion must be that in-school assessment should include the testing of the ability to measure, to use appropriate units and to employ appropriate measuring strategies in the investigative contexts of complex tasks rather than rely on assessments based on disembedded atomistic tasks alone.

7

Making and interpreting observations

7.1 Introduction, background and history

In presenting a review of this category of scientific activity it is worthwhile restating the view of the science teams relating to the nature of scientific observation as expressed in the equivalent chapter of the first age 15 report (Driver *et al*, 1982):

> 'The importance of observation in science has led to a considerable amount of philosophical reflection on what the act of observing involves. In the context of science, observation is not restricted to the sense of sight. It includes all the senses, and it is not a passive activity. Merely "seeing" something could be adequately tested by employing the methods used by opticians. When engaged in scientific observation the observer uses an intellectual framework in order to know what to see; how to look at the world. This framework is acquired through experience and training in the discipline of science. Ideas and theoretical notions about the world lead to the creation of a model which informs about what to select as noteworthy: what to disregard as spurious. Unfamiliar objects and events are interpreted in the light of what is new compared to what is already accepted. This process of analogy is fundamental to the use of models. The possession and use of the particular types of model found in science distinguishes scientific from other types of observation, such as artistic "appreciation" guided by aesthetic principles. It is considerations such as these that have led to the current view that scientific observation is "theory-laden".'

The development of this category in assessment terms through the questions and the marking methods employed has reflected this theory-dependent view.

The early deliberations which established **Observation** within the APU science framework as a discrete and practically assessed activity standing free from **Interpreting presented information** and **Applying science concepts** can be traced through the APU Consultative Paper (DES, 1977) and the Progress Report (DES, 1979a). Subsequent history resides in detail in the reports of the various surveys. Observation questions were allocated to three subcategories as described below:

Using a branching key
- pupils are required to identify objects using a branching key.

Observing similarities and differences
- pupils are required to make comparisons either between two or more objects or events, or between successive stages of an event. They may have to decide what is important and relevant in the comparison and record their decisions in a variety of ways.

Interpreting observations
- pupils are required to collect information by observation, often by making comparisons or noting changes, and then use what they take to be relevant to go beyond their observations to find patterns, explain what they observe, or predict.

Selections from each of the subcategories were used in the surveys of 1980 and 1981. **Making and interpreting observations** was not included in the age 15 monitoring of 1982, being surveyed again in both 1983 and 1984.

The period between the surveys of 1981 and 1983 enabled a thorough evaluation of the category to be made. Questions, mark schemes and practical resources were reviewed in the light of two years' pupil response data; in effect the first data to be gathered on questions designed to test the scientific activity of observation.

The second and third of the subcategories described above had been separated for the purposes of assessment in order to distinguish between the ability of pupils to 'make observations' and to 'interpret observations'. The essence of the difference resided in the *explicit* requirement in the latter for pupils to go beyond the immediate sensory information to make explanations, predictions or to state patterns.

The distinction between 'making' and 'interpreting' is arguably artificial and becomes problematic when scoring responses. In any case the view of scientific observation adopted by the teams (quoted above) has always required pupils to interpret and decide on relevance before making any response, regardless of whether the task has asked them to *record* their reasons. Interpretation had been rewarded in 'interpreting observations', but its role had been uncertain and unquantifiable in 'observing similarities and differences'. In consequence the two subcategories have been merged (although the various question types have remained discrete) so that a single subcategory score can be computed which encompasses interpretation at whatever level.

At the same time (between 1981 and 1983) pupil performance on 'using a branching key' was scrutinised. The result of this review was presented in the report of the 1983 survey at age 15 (Welford *et al*, 1986). The conclusion reached was that for older pupils (13 and 15) the main burden of questions calling for the identification of objects using keys was neither on observation nor on the logic of use of keys but to do with the meaning and interpretation of the words used in dichotomous statements. In fact the level of observation was often trivial – eg 'is it brown?' or 'is it green?'. While it is true to say that keys are applications of observation, what they measure in terms of assessing competence 'to observe' is unclear, and their precise and quantifiable contribution as a component of 'scientific observation ability' is uncertain. In the light of these considerations it was decided not to include key questions within the pool of questions for testing **Making and interpreting observations**.

Keys are still, however, within the pool of observation questions at age 11. The question format has been changed for 11 year olds from the binomial presentation to a more pictorial branching outline. In this way the logic and language burden is reduced and it is argued that observations which they have to make are not trivial for younger pupils. For the 1980 and 1981 surveys at age 15 there were questions in the pool which required pupils to make drawings of objects. Again, in the revision of the question types, the drawing questions were removed at ages 13 and 15, part of the rationale being that there are some important features of objects which cannot adequately be described by such a response on its own; a combination of words *and* diagrams being of greater value in describing the article. It is also true to say that few valid questions could be set from content areas outside Biology, and so the aim was not realised of assessing pupils' competence across a wide range of content and contexts to see whether pupils would deploy their skill generally.

Although the question banks at the three ages developed separately the monitoring teams had always sought to produce items which were common to more than one age. However, small differences had arisen (for instance, in precise wording of questions, weighting within mark schemes, presentation of resource etc) usually quite intentionally to reflect differences in scientific experience of pupils at the three ages. Such differences associated with the questions proved large enough to prevent direct and valid comparison of performance across the ages. The final rationalisation before the 1984 monitoring reviewed these age specific variations and resulted in a common question pool operable at ages 13 and 15, with some small overlap with that at age 11. This merger was not intended to deny the existence of age specific differences in applying the skills of observation. Indeed differences in performance between 13 year old and 15 year old pupils can now be more confidently ascribed to the widened experiences enjoyed by the majority of the older pupils.

Rationalisation of mark schemes has also taken place to standardise the award of marks within question types and a common system of categorisation of pupils' responses has been used where possible.

7.2 Test administration and question selection

The random selection of questions from a large pool representing the domain of a category has been described previously in Chapter 4 of this report. Of the three categories assessed in practical mode **Making and interpreting observations** is unique in that it is domain-sampled in this manner. Survey cost constraints have limited the number of questions used to a maximum of 45 in any one survey. Tests are administered utilising the circus format whereby a group of pupils attempt a series of experiments individually moving from one laboratory station to another at timed intervals. In 1984 three circuses were used, each comprising 15 questions arranged at nine stations.

Each banked question has been extensively trialled to determine its characteristics prior to inclusion; one objective being to determine the likely length of time of completion by the majority of pupils. As a result two (or even three) shorter questions may be attempted at some circus stations; the longer questions occurring singly.

As described in the previous chapter for the subcategory **Use of apparatus and measuring instruments** pupils have been assessed in groups of nine, each individual working around a circus of experiments moving from one station to another in a set sequence at intervals of eight minutes. The tests were set up and run by trained testers/supervisors who were, for the most part, experienced Heads of Science. These supervisors have taken the standardised sets of equipment to each survey school. Further details of recruitment and training of teachers as supervisors, and of the provision of equipment are given in the Science Report for Teachers: 6 (Welford *et al*, 1985).

7.3 Pupil performance on 'Making and interpreting observations'

This section starts by reviewing performance on **Making and interpreting observations** as a whole before going on to look at characteristics of performance on questions of various demand types, set in different contexts, or employing different types of apparatus and resources. For each question type there has been an attempt to look at characteristic errors, and the performance of pupils of differing ability is compared.

As with the other subcategories discussed in Chapter 4 the number of questions available for selection in 1980 was relatively low. Between 1980 and 1983 the pool increased in size, only to diminish before the last survey with the exclusion of the two question types – keys and drawings – for the reasons already discussed.

Mean scores were computed in 1980 and 1981 only for the subcategory **Observing similarities and differences.** The subcategory mean score estimates calculated in 1983 and 1984 were based on the whole selection of questions which represented the domain. Such changes, allied to those in question pool size and composition, preclude sensible comparison between surveys or discussion of patterns of performance since 1980. It is worthy of note, however, that the scores for those questions used more than once since 1980 (and marked to the same criteria) have shown very little variation from year to year.

The subcategory mean score increased slightly from 46 per cent in 1983 to 50 per cent in 1984 which might be due (partially at least) to the deliberate exclusion of certain question types but is far more likely to be a reflection of all the rationalization policies again referred to previously. A significant number of questions formally administered only at age 13 were by now part of the common 13/15 question pool, and there were questions lost which had been unique to age 15. If the former may be assumed to be easier, and the latter more difficult, then the net effect of gain and loss will have been to elevate the mean score of 15 year olds. The depression of mean score observed at age 13 serves to confirm this effect (Schofield *et al*, 1988).

Although comparison between years is not sensible due to the perturbations already discussed, it is worthy of note that there has always been a slight though statistically insignificant performance gap in favour of girls; a gap moreover which is detectable in 11 year old pupils, and persists through the 13 year olds into those aged 15. The difference reached statistical significance in 1984 and it will be interesting to see whether it reappears in the next survey in 1989.

Table 7.1 shows the main groups of question types represented in the pool of **Observation** questions, and indicates the number of each type used in 1984. Question type mean scores are given, with the ranges of individual question mean scores for each question type shown in parentheses.

Detailed comparison between 1984 results shown in Table 7.1 and those of previous years is not fruitful because of the changes brought about by the rationalization of the 13/15 question pool. However, two factors noted for earlier surveys are again evident here.

Firstly, while there are differences in mean scores between the various question types the range of individual question mean scores *within* any question type is much greater.

Table 7.1 *'Making and interpreting observations' – summary of mean percentage scores on different types of question (1984 survey)*

Question type	Number of questions	% Mean score: All pupils (range)	Boys	Girls
1 Given objects or photographs, group objects into self defined classes, or identify the rules used to classify the objects, and add further objects to classes	6	39 (17–60)	38	40
2 Given objects, photographs or events, describe similarities and differences	12	53 (24–75)	51	54
3 Given an event, make a record of change	3	43 (37–54)	40	45
4 Given an object, select the matching drawing from a range of drawings	5	61 (31–77)	58	63
5 Given events, make a record of observations and either give or select appropriate explanation	3	37 (19–46)	36	38
6 Given objects, photographs or events, make a note of differences and make or select a prediction consistent with the observed data	8	51 (31–85)	52	50
7 Given events, make a record of changes, and either make a prediction consistent with the data or identify a pattern in the observed changes	8	52 (22–85)	51	53

Secondly, the differences in performance between the sexes detailed in previous years is again apparent with girls doing better on nearly all the different types of question. As in 1983, if the question demand is on the description of objects, photographs or events then the girls are almost invariably better. Both question types 2 and 3 in the table require descriptive writing and girls outscored the boys on 12 of the 15 questions (ranging from one to seven per cent better) – there being no difference on the other three questions. The 15 questions ranged across a variety of content areas.

The picture relating to gender differences for all the other question types included in this category is not as clear. For the two groups just outlined it may be that the descriptive demand of the questions would seem to favour girls, but no such homogeneity is detectable between or within the other types of question, where individual question scores were much more variable between the sexes. Indeed a question's content would seem to be a strong influence on gender related performance differences. For instance, boys outscored girls by up to eight per cent on questions involving electricity, whereas girls did seven to eight per cent better when classifying chemicals, classifying everyday materials, and when identifying patterns of size, shape and number of soap bubbles being blown through different frames.

Various studies of the affective influences on performance (eg Smail, 1983; 1984) have suggested that girls find certain science contents more to their liking. These include mineralogy and chromatography, and the results of all **Observation** questions used to date at age 15 in these contents support such a contention: girls have always outscored their male peers. However, the suggested interest of girls in anatomy and physiology does not evidence itself in increased performance in this category. Questions about skulls, teeth and alimentary canals have resulted in identical scores for the sexes or slight differences in favour of boys.

7.4 Classifying

The report of the 1981 survey (Driver *et al*, 1984) described performance in detail for a question where pupils were asked to put a set of 20 photographs of faces into two mutually exclusive groups, to state the basis for the grouping used and then to repeat the exercise making a different grouping. A worked example of such a classification based on gender was completed for the pupils to show what was required. About 30 per cent of pupils were able to form two different and mutually exclusive groupings on *both* occasions (45 per cent and 41 per cent successful on the first and second occasions respectively).

Seven different questions of this type have been included in the surveys since 1981 (two of them in successive years with almost identical results). While there is no consistent pattern of difference between the sexes, there are some trends (of a non gender related nature) of note. A far larger proportion of pupils were actually able to classify or group objects (plants, bones, chemicals, everyday solids) than were able to articulate their grouping criteria in writing. For instance, in one question pupils were given photographs of nine different plants sorted into three different groups. They were asked in part (a) to identify the grouping criteria, in part (b) to allocate a living plant specimen to one of the groups, and in (c) to say why they had classified the plant thus. Between 25 per cent and 40 per cent of pupils were able to describe the grouping criteria for each of the three groupings in part (a). However, although 59 per cent allocated the specimen to the correct groups in part (b) only 39 per cent scored *any* marks in part (c) for saying what features they had used to choose between the groups. Interestingly, while there was no difference between the performance of boys and girls in parts (a) and (b) girls outscored boys in part (c) which incorporated an extended writing demand.

The gap between the least able pupils and most able pupils was less marked when they were asked *to group* than it was when they were asked *to describe* grouping criteria. On the plant grouping example above 69 per cent of the most able and 43 per cent of the least able correctly allocated the plant specimen to its group, with 50 per cent and 22 per cent respectively giving adequate explanations for their choice. A similar pattern was noted for all the classification questions so far used.

7.5 Describing similarities and differences

A large number of questions have been used (question types 2 and 3 in Table 7.1) in this area. There is a noticeable imbalance in the **Making and interpreting observations** question pool between describing static objects or photographs (20 different questions used) and describing events (five different question used) and between the contents employed by the two groups of questions. Only a small proportion of group 2 questions are set in the context of physical science and none of the group 3 questions involve biological science. It is not hard to see how such a situation has arisen. Firstly description of, say, a stationary pendulum serves little purpose since its noteworthiness is connected with its motion. Secondly, constraints of large scale surveying across three countries, with testers visiting many different schools, make the unpredictable, inconsistent and slow response of living materials poor subjects for questions lasting a maximum of eight minutes.

However, notwithstanding such imbalances the descriptive demands posed by the questions nearly always resulted in girls outscoring the boys.

Limits to scoring on questions requiring pupils to generate the attributes of objects or events have been suggested in past reports. The most obvious limit has been caused by the omission of observations rather than by the making of incorrect observations. Pupils rarely explored the full range of possibilities when describing similarities and differences, being satisfied with two or three observations when there were many more available for description and often when more than two or three were specifically requested.

When describing events (as with identifying variables in relationships or identifying classification criteria) performance may be seen to be limited by the tendency of pupils to concentrate on a single observation. This has been suggested as having its roots in the expectations and experience pupils have of this type of task (1983 survey, Welford *et al*, 1986). Pupils are probably used to directing their focus on a single variable in a relationship, since illustrative experiments in their classroom learning experience have been used to pinpoint the effect of, say, mass or length on pendulum periodicity. It could well be that pupils may even have been positively discouraged to gather data other than those central to the concept being studied, since the description of other variables may be seen to divert pupils and detract from their ability to understand the key relationship.

Conversely, in some circumstances it may well be that a single observation is sufficient, or even the most appropriate answer to a question. For instance it is highly likely that a single and correctly observed attribute of a cranefly which distinguishes it from both an ant and a wasp will satisfy the demand of a question expecting such differentiation (see Driver *et al*, 1984, p53). In fact analysis of pupils' responses to the insect question just described showed that single accurate statements were given most often by the most able pupils who were studying Biology. Similarly, in answer to questions set in a Physics context, the single but telling response was given more often than average by the able pupils studying Physics.

The description of similarities and differences between objects or between events has usually resulted in pupils making a larger number of observations than when describing relationships, or predicting variable values. However, such a feature does not mean to say that pupils do better when engaging in tasks of the former type than they do when attempting the latter. In fact were the hypothetical situation to arise when the score achieved for a response was arrived at by dividing the number of observations made by the number possible, then it would most certainly be the case that the first type of task would produce rather low scoring. As it is, most of the mark schemes for the similarities and differences questions credit up to ten discrete observations, and the limiting factor on scoring has been that pupils do not generate as many separate statements as requested although the number demanded by the question is often only a tiny fraction of those available.

It is interesting to comment on the effects of the curriculum background or experience of pupils for these same types of questions. A careful scrutiny of a subsample of scripts of a question requiring comparison between three beetles has shown that it was those studying Biology who made an orderly review of the type of insect characteristics usually useful in taxonomy. Those studying Physics on the other hand made fewer such statements, but features of shape, materials and support were more common in their observations. Obviously both types of observation are equally scoreworthy and so, especially at the upper end of the ability spectrum, the curriculum background of pupils has not influenced performance as measured. Such an argument does not, however, deny the theory-laden nature of observation, it merely suggests that since there are many perceptions of equivalent relevance, then the mark schemes have not operated in such a way as to exclude one perception in favour of another.

Findings at younger ages that pupils found differences easier to describe than similarities have not been consistently replicated for this older age group. A slightly higher level of performance was shown in 1981 for noting differences than for observing similarities. This tentative finding has not been confirmed in subsequent surveys; there being no difference at all in 1983 (Welford *et al*, 1986), or in 1984.

Driver *et al*, (1982) noted that some pupils were repetitive in describing similarities and differences, eg 'P is bigger than Q' and 'Q is smaller than P'. This has also been confirmed by successive surveys. Similarly it was noticeable that pupils may have located a particular attribute of an object or event but then fallen down when making the comparisons between the objects or events on view.

For instance pupils observed that a cranefly had one large pair of wings, but did not immediately compare the number of pairs of wings possessed by a wasp. After observations of other features many then made a separate and again not comparative statement that 'a wasp has two pairs of large wings'.

In ascribing features of performance to one group of pupils or another based on their examination entry potential or test performance, the picture again shows polarization. The errors described above are not made exclusively by pupils of the lower bands, but such pupils do make all the errors described more frequently. Their lack of descriptive language structures must put them at a disadvantage when faced with questions imposing descriptive requirements. The less able pupils were also much less likely to make their observations comparative. Having described an attribute for one object they often failed to describe that attribute for the other object(s).

Mode of resource presentation

Static 'similarities and differences' questions are constructed either around a photographic resource or based on real objects. Whether one or other mode is more effective in promoting responses has been one of the concerns of the teams. Photographs are generally cheaper, easier to administer in testing, and have the distinct advantage of being identical across all replicates. Objects possess the advantages (and disadvantages) associated with 'the real thing'.

Whether or not the mode of presenting a resource (photograph or object) has had an effect on pupil performance has not emerged from the results. For instance, in 1983, a question based on two sheep's vertebrae had the same mean score as a second question of very similar demand based on photographs of three beetles (both 24 per cent). A third which presented three specimens of different species of thorn moth for description had a score of 76 per cent in contrast with a score of 44 per cent achieved on a question based on photographs of insects.

A limited number of exploratory questions were used in 1984 whereby exactly the same resource was presented in two modes – one version using objects, the other high quality photographs of the same objects.

In performance terms the issue remains unresolved. A question which presented pupils with two large stag beetles of different species to compare resulted in a mean score of 52 per cent. The photographic version of the beetles questions had a mean score of 24 per cent. Actual bones to be described produced a score of 24 per cent compared with a score of 36 per cent achieved for the question based on a photograph of bones. Given seeds to handle and to describe similarities and differences pupils scored 44 per cent. In answer to exactly the same question based on photographs of the seeds the mean score was 38 per cent.

The results of this small scale exploration are singularly inconclusive – in some questions the objects were more successfully described, and in others their photographs would seem to have resulted in higher scoring responses. Maybe in some cases the photographs served to highlight certain features, and in doing so reduced the range of confusing attributes to consider. In others it may be that two dimensional representations restricted the possibilities for observation, and so reduced the scoring opportunities for pupils.

7.6 Making or selecting explanations of observations of events

The number of questions of this type which have been used in surveys so far is limited (five different questions in 1983 and 1984) and so generalizations are tentative at best. However, the science experience of pupils would seem (not unreasonably) to have a bearing on their ability to explain the observed data. An example might serve to illustrate this point. A question (similar to Example 4.8 in the 1981 survey report [Driver *et al*, p55]) faced pupils with two identical tins of the same mass hanging from string of the same length. One tin was filled with sand, the other with low viscosity fluid. Pupils were required to spin and release the tins to observe differences in their motion, and then to explain their observations. There was little overall variation in the performance of pupils of varying curriculum background when noting the differences in motion of the tins, but pupils of 6+ O-level ability studying Physics scored one-fifth more marks for their explanations than did those of equal ability without a recent Physics background. The physicists related the observed motion of the tins to the contents with much greater fluency and accuracy. Another question requiring pupils to select the best explanation for observations relating to the evaporation of liquid favoured those with a background in physical science: observations of differential expansion of glass, polythene and water likewise drew more knowledgeable responses from those studying Physics and/or Chemistry.

The five questions in this group show patterns of performance related to academic ability which are quite similar to those described in Chapter 9 **Applying science concepts**. Explanations of observed phenomena are far more accomplished from the above average pupils, with a marked decrease in performance from the bottom fifth of pupils.

7.7 Selecting or making predictions consistent with observed data or identifying patterns in observed changes

Questions of these types listed in Table 7.1 have been considered together in this review of performance. The variation in mean scores are similar for the two groups (31 per cent to 81 per cent and 22 per cent to 85 per cent) as are the group mean scores (51 per cent and 52 per cent).

In what was arguably the easiest question, pupils were required to put their fingers inside an oblong box to feel pieces of five different materials, which they had to identify. The question has been previously reported at both younger ages (Harlen *et al*, 1981; Schofield *et al*, 1986). Performance was very similar for 13 and 15 year olds with two-thirds of pupils correctly identifying all five materials, and around 90 per cent being correct for three of the five substances.

The question with the lowest mean score presented pupils with a chromatogram of five inks and asked what they could tell the questioner about the inks. Pupils were then required to predict the appearance of the chromatograph which might result from a mixture of two of the five inks. As described previously girls outscored boys (24 per cent to 18 per cent) with pupils of 6+ O-level ability studying Chemistry performing better than others of similar ability. In making deductions about the inks there was not as much difference in performance between pupils of different ability as there was when offering a prediction. Able Chemistry takers were fully 14 percentage points ahead of others of like ability (49 per cent to 35 per cent with a mean of 28 per cent on this part of the question). Four in five of the less able pupils failed to score when making this prediction. These two examples serve to give some indication of the range of performance characteristics across questions of these types and illustrate the question specificity of some features of performance. However, some more general comments are possible.

Identification of variables

Many of the questions grouped here required pupils to observe and describe relationships within dynamic systems, or between structures and their functions. The report of the 1983 survey (Welford *et al*, 1986) concluded that most pupils (more than 75 per cent) were able to identify the central dependent variable within a relationship. The majority (two-thirds) were able to identify the independent variable and to go on to describe the relationship between the variables. The conclusions were based on a single question analysed in depth, but have been confirmed by other analyses of similar questions.

If the pupil had to observe and to describe a single dependent – independent variable relationship, performance patterns followed those outlined above. However if the pupils were faced with more than one dependent variable changing as the independent variable was systematically varied, or with systems involving more than one independent variable, then performance was much less predictable and generally lower. For example, pupils were given concentrated soap solution and a range of wire frames varying in size and shape. They were asked to observe the number, size and shape of bubbles they blew through the frames, and asked about the relationships between frames and bubbles. Fifty-four per cent of 15 year olds related *size* of bubbles to *size* of frame, 39 per cent related *number* of bubbles to frame *size* and 32 per cent related *shape* of bubbles to *shape* of frame.

The number of variables (dependent or independent), and their confounding, affected pupil performance. While two-thirds of pupils described single dependent variable-independent variable relationships the proportion who described one of the other more complex relationships fell to between a half and a third of those attempting the question. A far smaller proportion (c 10 per cent) included all variables in a systematic review of, for example, size, shape and number of bubbles in relation to size and shape of frames.

Expression of variable relationships

The foregoing discussion has centred on whether or not pupils have been able to identify the relevant variables in a relationship; their expression of that relationship is now discussed.

The inclination of pupils to generalize from sets of such related data which they have generated by their observations has been examined for a limited number of questions. For relationships involving a single dependent variable, between 40 per cent and 50 per cent of 15 year olds have expressed the relationships which they describe in general terms such as 'the heavier the object the slower the apparatus swings'. This proportion dropped to between one-fifth and one-quarter of all pupils when the number of dependent variables was increased. Non-general, specific statements were often employed – 'the apparatus moved quickest with the lightest weight'; many pupils giving these specific statements often making more than one such description – ie adding 'the apparatus moved slowest with the heaviest weight'.

However, regardless of the number of variables involved in an experiment, 15 year old pupils who identified the relationship were at least as likely to couch a statement in general terms as to make a more specific observation statement.

Question content would seem to be a considerable influence on pupils' performance in these kinds of questions, and is related strongly to the taught science experiences of the pupils. For example, pupils studying Physics have scored considerably better on identifiably 'Physics' questions than have those of equal ability not studying the subject.

The academic ability of pupils also affects performance. Although the least able were not much less likely than their higher attaining peers to be able to pick out the variables in a single dependent – independent variable relationships, they were *much* less likely to use the power of generalization in their descriptions of the effect of one variable on another.

Performance on questions about structure-function relationships were discussed in detail after the 1983 survey (Welford *et al*, 1986). It was suggested that such questions are highly content- or theory-laden, and the evidence presented suggested that pupils studying a relevant science subject were at a distinct advantage; advantage increasing with amount of science studied. Results from the 1984 survey support these findings; pupils of 6+ O-level ability studying Biology consistently outscored their peers on questions about skulls/dentition and diet, birds' beaks and diet, birds' feet and habitat, and seeds and dispersal of seeds. The lowest attainment band performed much less well than the other 15 year olds on these questions, with the top 20 per cent showing much firmer grasp both of what observations were relevant to record, and how to relate observed structures to known functions.

7.8 Summary

A variety of types of question have been asked in the surveys of this practical activity. A review of the history of the question types has shown that some types have been rejected as inappropriate for testing observations at age 15. The remaining types have been described and performance data presented. The range of individual question mean scores within any one question type has been shown to be greater than the differences in group mean scores. For this reason detailed comparison of performance between 'question types' has not been attempted; no statements can be made which describe performance on one such categorisation as more or less difficult than any other.

Constraints are imposed by the demands of large scale surveying, which have had implications in terms of the range of content which it has been possible to include across all types of question, and for the feasibility of the use of certain apparatus and resources.

The view of scientific observation which has been developed by the teams and put into operation through the questions representing the domain has been described and illustrated throughout the chapter. Any discussion of performance characteristics of pupils must address itself to the variety of factors which influenced pupils in making their responses. Among these are the characteristics of the pupils themselves and their differing background and experience, as well as the nature of the questions in terms of their procedural demands and the level of conceptual understanding needed to answer the question. Question content and the nature of the marking criteria adopted overlap both with the above and with the underlying view of the nature of observation. Interpretation of performance is thus reflective of the interaction of many influences on responses and their assessment.

The better performance of girls on this category as a whole is repeated for most types of question and not just those which impose a higher descriptive demand.

Except in certain specific areas such as electricity and biological structure/function relationships girls were better than boys at **Making and interpreting observations**.

Responses to questions testing observation rely heavily on descriptive language and knowledge of specialist terms or phrases. While for the purposes of these tests every effort has been made not to disqualify an observation through lack of clear communication, many answers have been examined for the effectiveness of language used. The incidence of use of specialist terms and structures has been low, and to a certain although unquantified extent, must have limited pupils' capacity to describe or record their observations.

In connection with expression, especially in a scientific context, the ability to generalize data has been advanced as a sign of increasing sophisitication. It has been argued that younger pupils attempting **Interpreting presented information** questions were not good at generalizing about relationships (Schofield *et al*, 1988). It would appear that in **Making and interpreting observations** at least, 15 year olds did make generalizations with greater fluency, but they did not invariably describe relationships in general terms. It may be the case for these older pupils also that they either did not appreciate the power of a generalization in science, or that they have not experienced sufficient opportunities to know when (or how) to generalize their data.

Responses to the types of question requiring the description of similarities and differences have shown some qualitative differences relating to the science subjects being studied by pupils. Such differences have not affected scoring significantly since these mark schemes encompass different perceptions of relevance.

The other question types used in this category have, however, produced performance characteristics which show a far greater influence of the immediate curriculum background of pupils. Descriptions of relationships for instance have usually been based on dynamic experiments where the number of variables was more tightly constrained. The number of attributes of an inertia balance, for instance, which change as the load is increased are not infinite. It has become clear that most pupils have rarely given more than one such attribute in their answer. In an example such as the inertia balance a greater proportion of those pupils doing Physics have both identified the variables and described the relationship correctly. Those who did describe additional variables were the most able pupils studying the science subject most closely paralleled by the question content.

Mark schemes for such questions are more sharply focused than are those for the similarities and differences types. It has been noticeable, and not in any way surprising, that many more responses are incorrect; after all, with only a limited number of observations available the chances of being wrong are much greater.

However, most pupils (except the most able) did limit themselves to single observations when attempting these more interpretative tasks. This has happened even when the question has cued pupils to look for several relationships. Could it be that pupils doing class practicals have been used to being directed to look for a single relationship, and to look no further when they have identified the relevant variables? After all, much practical work even of an investigative nature in the fourth and fifth form is tied closely to the acquisition of conceptual knowledge. To look for other variables may obscure the central point of the lesson or topic area. However, if the development of procedural knowledge is to feature in future schemes of work, then perhaps such directed experimentation will need to be supplemented by less constrained observation.

Errors made which might be described as characteristic of the least able, the average or the most able pupil are not starkly evident. Less able pupils were, however, more likely to make too few observations and less likely to make a systematic review of the objects or events under consideration. As a result their responses showed a tendency not to make their descriptions comparative when describing similarities and differences. Such pupils were much less likely to explain observations or to make explicit their interpretations of their observations. This may have been either because they could not (lacking the specific concepts and/or the necessary language skills) or because their perceptions of the relevance of their observations were more varied. Least able pupils, while usually capable of identifying the variables in a simple relationship, were very much more sensitive to the effect of increasing the complexities of a relationship. They were also less likely to be able to generalize a variable relationship or to make accurate predictions based on observing patterns.

The theory-laden view of scientific observation has been well supported by the evidence of the influence of the curriculum background of the pupils on their performance.

Thus while **Making and interpreting observations** is included for testing in the APU science framework of scientific activity categories, it may well be that the appropriate place for its specific inclusion in taught science is a practical test closely related to the pupils' conceptual knowledge base. In fact the dual development of conceptual knowledge and procedural competence may well be achieved through a coherent strategy of learning and teaching focused through **Observation** tasks in scientific practical and investigatory work.

8

Interpreting presented information

8.1 Introduction

The question of the boundaries between **Interpreting presented information** and other categories and subcategories is among the most difficult in the assessment framework. On the one hand, since most of the data are presented in standard forms it is necessary to establish the boundary with **Using graphs, tables and charts**, where the extraction and representation of data is the main focus. Similarly it is necessary to distinguish it from **Application of science concepts** where the understanding of taught science dealing with specific phenomena is held to be a prerequisite for successful performance in handling given information.

The question of these 'external' boundaries of the subcategory must, ultimately, be a judgemental matter. It is possible to define boundary conditions, but the application of these will never be entirely automatic. The boundary with more routine extraction or representation of data has been determined by the need to use some generalized relationship in moving from the presented data to the required response. In **Using graphs, tables and charts** such a generalized relationship is not implied or required. Some questions, such as those involving rates of change, sit awkwardly on this boundary and have been allocated to **Using graphs, tables and charts**. The boundary with **Application of science concepts** is more difficult. All questions require some minimal conceptualization of the phenomena under discussion. In general this increases as the data shift from quantitative to qualitative forms. The boundary criterion which is used in this case is therefore based on the apparent *necessity* to use (not the mere possibility of using) ideas which a pupil is unlikely to have met outside science lessons.

Two major components of interpretation were allocated to distinct subcategories. These were: **Judging the applicability of given generalizations** and **Distinguishing degrees of inference**. Questions of the former type were not easily distinguished from fixed response questions requiring pupils to generalize from given data. The latter subcategory was given limited attention, after it was found to be difficult to set up questions in which the 'degree of inference' required varied unambiguously. These two subcategories were therefore not used to generate scores, and this review focuses on interpretation proper rather than these subordinate activities.

It was established at an early stage that it was necessary to specify the character of questions more precisely than did the subcategory title. Interpretation, when not requiring the deployment of taught science concepts, was taken to involve the following general tasks:

(a) **prediction**: prediction of a new datum or sequence of data. At its most routine this involves extrapolation or interpolation of the given data points. It may involve the perception of a generalized variable relationship followed by deductive reasoning. In the case of qualitative data it is frequently necessary to deploy some ideas concerning the specific phenomena involved. What constitutes a datum in questions of this type is determined by the framework of the question. There has not been an attempt to distinguish questions based on 'observable data' from those based on derived or theoretical knowledge.

(b) **description/generalization**: describing relationships inherent in given data, often involving either deciding which of a number of variables is effective in altering a given variable, or identifying some general characteristic of the relationship between two given variables. In many cases this corresponds with inductive reasoning.

(c) **justification of a prediction**: this occurs only in cases where a prediction has been made, and involves the pupil in giving an account of how the prediction was obtained. It involves not merely the expression of some generalized relation derived from the data, but also some indication of *how* that generalization was applied in the specific case involved.

In combination with each other and various response modes these tasks have been used to generate eight question types. These are shown in Table 8.1 (p65).

Pupils' performances on this pool of questions generates an overall subcategory score which is reported and analysed for various populations in Chapters 4 and 13. It is not possible directly to interpret the subcategory scores in terms of specific pupil competencies. However, response data *can* be made more interpretable and this will be undertaken in the remaining two sections of this chapter. Firstly, scores can be broken down according to the various classes of tasks which have been identified above in the hope of observing informative *general* performance differences. Secondly, pupils' responses to individual questions can be categorized, in order to identify classes of response which are informative about the *origins* of poor performance.

Table 8.1 *The eight question types used in 'Interpreting presented information'*

Question type	Task* and response mode**	Percentage of pool
1	DG	18
2	DG + PG	8
3	DG + PS	7
4	PG	21
5	PS	15
6	DS	14
7	PG + JG	9
8	PS + JG	8

*D = Describe/generalize **G = Generate (prose/short answer)
*P = Predict **S = Select (fixed response)
*J = Justify

8.2 The relation of mean scores to class of 'Interpretation task'

The overall percentage mean score for the questions present in this subcategory which have been used in surveying is 44 per cent. This section surveys overall mean scores on the classes of question identified above, ie generalization from data, prediction of new data and justification of such predictions. The impact of ability on the structure of performance will also be discussed. This will be based on the relative performances of the five ability groups defined in terms of pupils' scores on this subcategory (representing in each case approximately one-fifth of the full ability range).

Performance data on the two major classes of tasks are shown in Table 8.2. These data include parts of questions where the scores for these can be separated.

Table 8.2 *Performance on questions involving prediction and generalizing from/describing data*

Task	Percentage mean score	Number of questions	Maximum score	Minimum score
Predict	48	73	92	13
Describe/generalize	47	46	86	14

While it might be expected that the task of prediction subsumes that of generalization, and is therefore likely to generate lower performance levels, Table 8.2 shows that there is no significant difference between pupils' performance on the two classes of task. Variation of mean scores *within* each class is very much greater than that across them. One must however be wary of comparing mean scores across these two classes of tasks. Differences in mark schemes (eg such as the fact that those for predictive tasks are much more dichotomous) are important. Table 8.3 shows performance on these two classes across the five ability groups.

Table 8.3 *Scores of performance groups on prediction and generalization tasks*
(Percentage mean score)

Performance group	Prediction	Description/ generalization
Most able	74	75
Above average	59	60
Average	49	47
Below average	37	34
Least able	20	17

As is often the case, the fall in performance is quite uniform across the types of tasks undertaken. In general only deviations from uniformity will be referred to, and the full data on the performance groups will not always be given.

Response mode has an important influence on overall mean score. This is partly perhaps through the influence of the reduction in the non-response rate (for the less able) in fixed response questions. However, it is also the case that these questions direct less able pupils into the broad type of response which is expected, and that this eliminates one difficulty such pupils face. It will be seen later that the responses of the less able to more open-ended questions frequently indicate a complete failure to perceive the demands of the question congruently with the intention of the question writer.

Those tasks involving prediction appear in two forms:

- prediction alone (question types 4 and 5);
- prediction together with either an explanation (justification) or prior generalization (question types 2, 3, 7 and 8).

It is possible in many cases to separate the scores for these two parts of the question, and thus to observe the effect of this extra requirement on pupils' performance in prediction. Data on these two distinct types of prediction are shown in Table 8.4.

Table 8.4 *Mean scores for prediction alone and for prediction with explanation/generalization*

Task	Percentage mean score	Number of questions	Maximum score	Minimum score
Predict only	44	48	92	14
Predict *and* generalize or explain	55	25	86	13

Table 8.4 indicates an improvement in performance on predictive tasks when pupils are asked to undertake this extra activity. A possible explanation of this again is that pupils, and particularly those who are less able, are cued more effectively into the *requirement* to *use* the given data in specific ways. An analysis of these data by performance group is shown in Table 8.5 (p66).

Table 8.5 *Mean scores for prediction alone and for prediction with explanation/generalization according to performance group*

Performance group	Prediction only (A)	Prediction with explanation/ generalization (B)	Ratio $\frac{B}{A}$
Most able	70	80	1.14
Above average	56	67	1.20
Average	45	59	1.3
Below average	33	45	1.36
Least able	17	25	1.47

Though the effect is, as usual, rather small, it appears that the scores of the less able are improved in relative terms more than are those of able pupils when the extra demand is made.

Further analysis within the groups of questions identified shows that the lowest overall mean score (35 per cent) is that observed for that part of the small group of 12 questions where pupils are required to justify the prediction they have made (and for which a separate figure can be extracted). This is the most complex sequence of operations within the tasks referred to here, requiring both inductive and deductive reasoning to be made explicit. It is of interest that the largest overall mean score (64 per cent) is that for the 'prediction' components of the same questions. The score of the least able on this type of prediction is 32 per cent and thus, in relative terms, at its largest within these analyses – almost 40 per cent of the score of the most able. This is despite the fact that the actual justifications produced by the least able in these questions are extremely weak, with a mean score of eight per cent.

Data such as those shown illustrate the difficulty of attempting to draw generalized conclusions about collective performance on broad domains. The details of response are strongly affected by the characteristics of given questions and pupils' responses to them. Nevertheless it appears that increased involvement with given information, even when not itself readily articulated by pupils, can influence performance in the handling of such data for specific purposes. It seems also that this effect is most noticeable when the greatest demand is made on the least able. The pattern running through these data is that less able pupils are able to display relatively and absolutely improved competence when cueing reduces major barriers as to what the question intends (without, of course, demonstrating the *specific* skills required).

8.3 The characteristics of pupils' responses in 'Interpreting presented information'

There is much variation of response within each of the classes of question so far identified. In this section pupils' performances on specific questions requiring the interpretation of presented information are analysed.

Despite the variations across questions this analysis indicates that some generalizations can be made and that the causes of poor performance vary considerably across performance groups.

Generalization from given data in free response questions

In free response questions complex mark schemes are required if categories of response rather than mere scores are to be recorded. Only limited numbers of questions are marked in this way, and the following account is based on a sample of these.

Where the request is to relate two fairly well defined variables, the most able show some consistency across a range of questions. In most questions approximately 90 per cent expressed the relation. A typical breakdown of responses is shown in Table 8.6. The question to which these data refer, 'Indcomp (1)' (Figure 8.1) presented tabular quantitative data on energy use and agricultural employment for several countries.

Table 8.6 *Responses to 'Indcomp (1)' by performance group*

(Percentage of performance group giving indicated response)

Performance group	Attempt at generalization	Only specific data points given	Other	No response
Most able	88	9	0	2
Above average	69	11	14	4
Average	51	14	22	11
Below average	23	9	38	29
Least able	7	5	37	52

Figure 8.1 *'Indcomp (1)'*

Records are kept of the total amount of energy a country uses each year (from oil, coal, etc). From this the average amount used by each person in that country can be worked out.

The table below shows the percentage of the population working in agriculture and the amount of energy used per person each year for six different countries.

Country	% of population working in agriculture	Amount of energy used per person (units of energy per year)
Ceylon	50	0.8
Cuba	42	5.1
France	26	19.5
Italy	31	8.4
USA	12	66.0
W. Germany	23	26.8

Describe what the table shows about the way the percentage of people working in agriculture relates to the amount of energy used per person in a country.

....

....

....

....

....

....

Table 8.6 indicates that the remainder of the most able pupils and some from the middle performance groups tend to respond in terms of individual data points. It is only possible to speculate at this stage on the origins of this type of response. It may indicate that the pupil *can only* perceive specific data points in the information. Alternatively the pupil may not realise that a *generalized* response is required.

Table 8.6 indicates also that the lowest performance groups are responding differently. Citing specific instances is a valid utilization (if not *interpretation*) of the presented information. However, the types of response associated with the least able (and in some cases found as a supplementary component of the answers of the more able) under the heading 'Other' often involve a much wider interpretation of the question's demand. Pupils inferred characteristics of the countries involved which went beyond the data presented, often quite sensibly. They may have referred, for example, to the climate or techniques of agriculture. This can be seen as an alternative interpretation of the question demand. The information is taken to be merely a trigger for pupils' responses, rather than requiring manipulation to generate those responses. This reflects some of the points made in section 8.2. The performance of the least able is of course dominated by the large non-response rate. The origins of this and the capabilities of the pupils involved are not clear. Some insight into the latter might be gained from considering fixed response questions, which generally have much lower non-response rates, and will be discussed below.

The profile of responses observed for 'Indcomp (1)' can be compared with that for 'Netball (2)' shown in Table 8.7. This question also gives numerical data, based on the three variables shadow length and times of sunrise and sunset. (See Gott *et al*, 1985, Example 5.1, for the data given.) Pupils were asked to describe the pattern in the data. The more complex three-variable data allowed a greater diversity among the attempts at generalization, and these are indicated in the table.

In 'Netball (2)' pupils were cued more explicitly into the requirement to look for a 'pattern' in the data ('What pattern do you see in these results?'), and the data themselves did not have the wider social and practical relevance found in 'Indcomp (1)'. The impact of these changes is clearest among below average and least able pupils, who now focus more narrowly on the data, though apparently with little success. The tendency to cite specific instances is again most common among middle ability groups.

One final question of this type can be compared to those above to illustrate the elements of stability and variation in performance. In 'Evaporation' (Figure 8.2, p68) pupils were asked explicitly to relate crystal size and time for crystallization from a prose and pictorial description in which salt solution was left in different everyday environments. The shift to pictorial data and still greater cueing appears to produce substantial improvements in performance among the less able. The responses are shown in Table 8.8 (p69).

There is a sense in which performance among the least able improves across these three questions, although this is not altogether reflected in the mean scores obtained. Their responses clearly demonstrate a greater engagement with the anticipated question demand: approximately half the group attempt a *generalization* of the kind expected in 'Evaporation'. In addition the non-response rate shows a corresponding fall. Thus pupils are demonstrating an increased awareness of what the question is requiring them to do. Of course not all of these pupils have the skills to carry through this requirement. Nevertheless these data indicate that a major burden for many less able pupils is not only interpreting data but also interpreting the demand of more open-ended questions.

This type of failure represents only one aspect of poor performance and others can be identified. The tendency to refer to specific data points rather than generalizing explicitly has already been noted: it represents an alternative misjudgement of the question demand. Where qualitative variables are present the opportunity for misinterpretation of the intended variables is usually greater, and Table 8.8 shows that this can cause considerable problems outside the most able group. It is comparatively rare for pupils to fail in ways clearly due to lack of technical skills at reading specific data points except when working with complex data. The way in which these difficulties are manifested in particular questions and their relative representation among different ability groups, is diverse even among quite routine questions.

Table 8.7 *Responses to 'Netball (2)' by performance group*
(Percentage of performance group giving indicated response)

Performance group	Attempt at generalization				Only specific instances given	Other	No response
	Total	Correct	Incorrect	Unexpected			
Most able	93	88	1	4	6	0	0
Above average	80	64	5	11	13	7	1
Average	67	45	10	12	14	20	1
Below average	39	17	11	11	10	27	23
Least able	22	1	14	7	2	33	43

Figure 8.2 *'Evaporation'*

Pamela, Jean and Deborah had a solution of salty water which they divided into three equal parts and put into three dishes all the same.

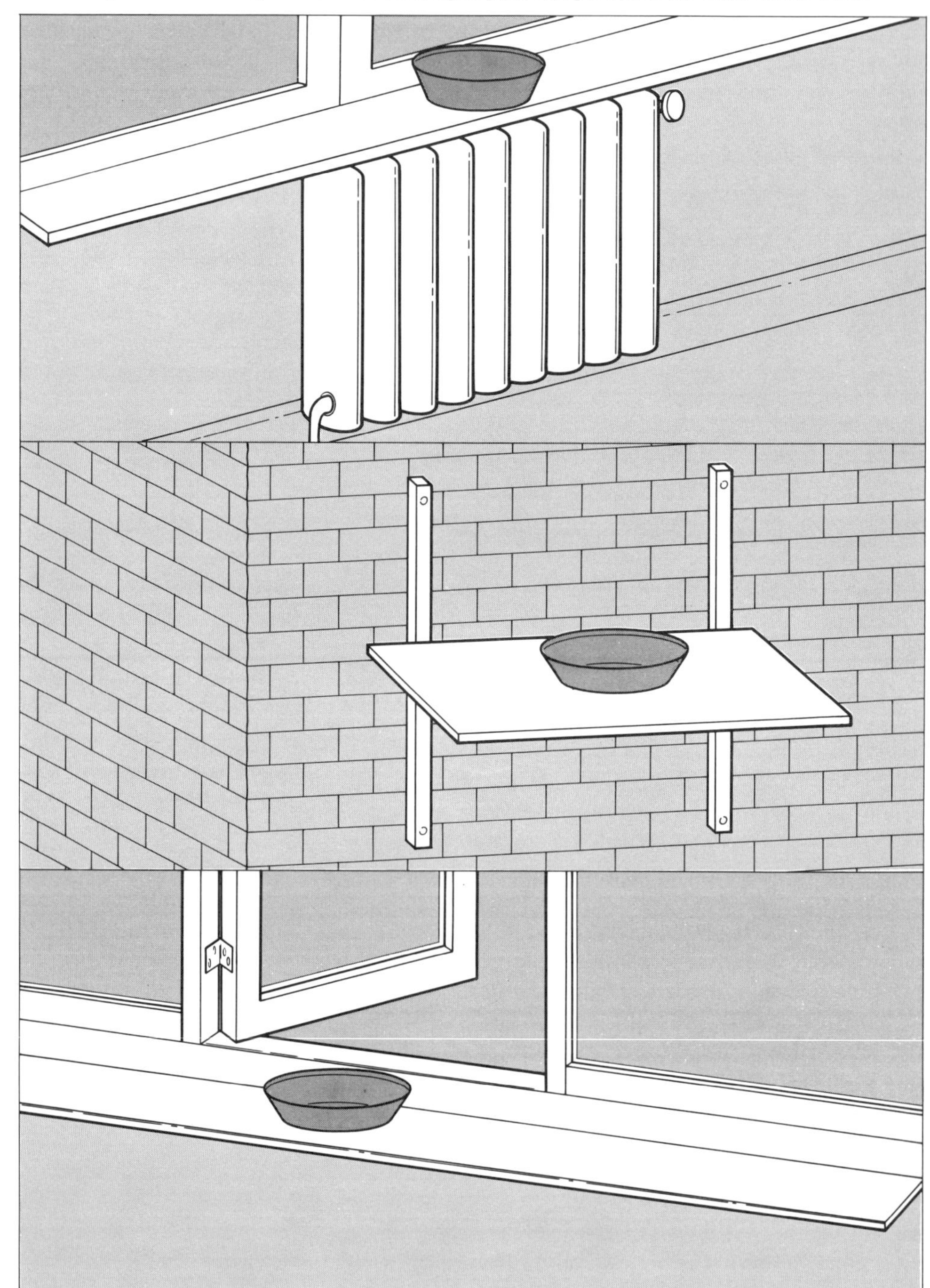

Pamela put hers on the radiator. The water dried up in *1 day*, leaving *powdery salt* (made of very small crystals).

Jean put hers on a shelf in a cool place. The water dried up in *7 days*, leaving *large crystals* of salt.

Deborah put hers by the open window. The water dried up in *4 days*, leaving *small crystals* of salt.

How does the time taken for the solution to dry affect the size of the crystals?

....

....

....

....

Table 8.8 *Responses to 'Evaporation' by performance group*
(Percentage of performance group giving indicated response)

Performance group	Attempt at generalization			Only specific instances given	Other	No response
	Total	Correct	Wrong or extra variable			
Most able	93	88	5	0	5	1
Above average	88	72	16	0	9	2
Average	79	61	18	1	16	4
Below average	64	39	25	3	19	13
Least able	45	22	23	4	31	21

However, it is of interest to note that the most able are rarely affected: their performance is, as a population, quite consistent. This indicates the extent to which they are sensitive to the often implicit requirements of the question writer, and resistant to distractions within the question itself. Thus in 'Indcomp (1)' the most able are much less likely than others to introduce 'irrelevant' information about the countries involved, even if, as seems likely, they have greater general knowledge. Similarly, in the later questions, the improvement in the tendency of the less able to generalize on the data presented (attributable to more explicit cueing and data less likely to trigger related ideas) often involves their using combinations of variables other than those signalled by the item stem. This was not observed to any significant extent in the most able. Thus the most able appear to detect and act on the implicit messages of the item stem in a much more sensitive, authoritative and 'decontextualized' manner than the less able.

Generalization from given data in fixed response questions

Fixed response questions are inherently of less interest in this section insofar as the possible errors are, so to speak, constructed by the question writer. Although the response rates among the least able are uniformly higher than in questions requiring self-generated responses, the pattern of their responses often shows an underlying random character. Performance on these questions does however appear to be improved by the fact that the responses available tend to focus on fairly well defined aspects of the data presented, removing the possibility of pupils introducing wide ranging misinterpretations.

The balance between responses is dominated by the incidence of characteristics which appear to draw pupils unable to perform the required operations from the correct responses and to those which are incorrect. This often depends on their concordance or otherwise with externally derived notions ('commonsense'). If incorrect options exist with a partial correspondence to complex data they also tend to prove attractive. It must be noted that many other characteristics (such as potentially ambiguous meanings for certain pupils) can play an important role. In illustration of these points Table 8.9 shows the distribution of responses of various types for the question 'Pop trends' (Figure 8.3, p70). This question presents GB data on population, birth rate and death rate for the period 1700–1950, and asked pupils to select a general conclusion.

The correct option (as the death rate decreased, the population rose) had no special characteristics in relation to the points referred to. Only among the least able is the response potentially representing a narrowly 'technical' error (E) quite common, confirming the impression noted earlier that simple technical incompetence in forms of data representation does not seem to have a key part to play here. Even among the least able the percentage selecting such responses is generally low, usually consistent with a repulsive effect superimposed on the tendency for responses to be randomized.

In this question the dominant incorrect response involves the commonsense notion that variation in birth rate is the key determinant of population changes, and this clearly proves a considerable attraction to the least able. The middle ability groups are surprisingly resistant to it. What is not clear of course is precisely how the least able treated the available data, if at all. These options involve clear and generalized contradictions with the available data. Among the possible reasons for their

Table 8.9 *Types of response chosen for the question 'Pop trends' by performance group*
(Percentage of performance group giving indicated response)

Performance group	Correct B	'Common sense' but contradicted by data A & C	Contradicted by specific data points E	Insufficiently precise D	No response etc
Most able	89	2	4	5	0
Above average	83	10	1	4	1
Average	62	25	1	4	8
Below average	55	26	9	5	4
Least able	25	42	16	5	11

Figure 8.3 *'Pop trends'*

The graph shows how the population of Great Britain changed between 1700 and 1950.

The bottom graph shows how the birth rate and death rate changed during the same period.

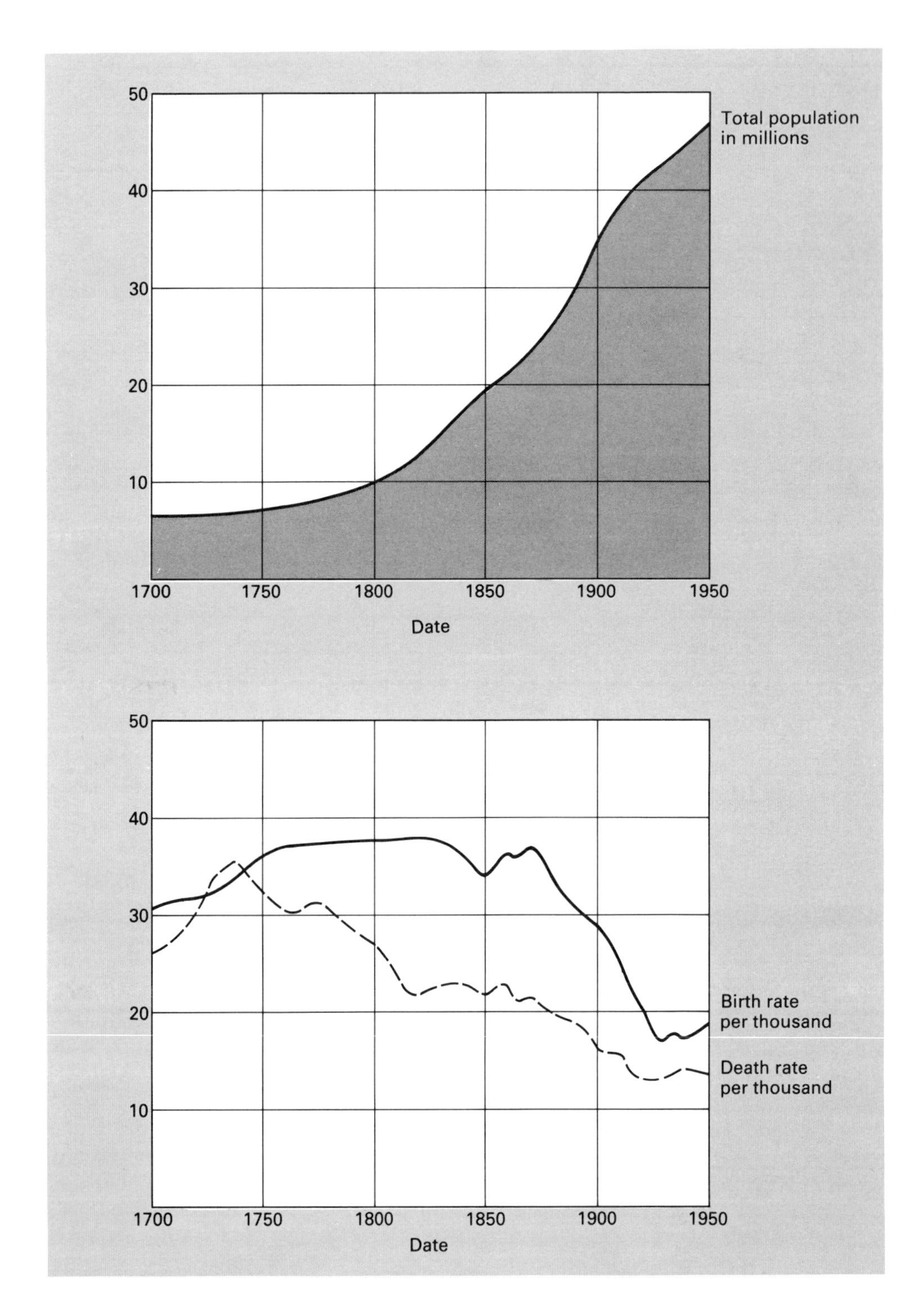

Look carefully at the graphs and then read the statements below about the changes between 1700 and 1950. Only one is true. Which is it? Put a tick in the box beside the one you choose.

☐ A As the birth rate decreased, the population fell.

☐ B As the death rate decreased, the population rose.

☐ C The higher the birth rate, the bigger the population.

☐ D The birth rate dropped twice as much as the death rate.

☐ E The death rate was always lower than the birth rate.

being selected two can be singled out. It may indicate that pupils do not understand that the question asks them to assess the *data* and the *statement* in relation to each other. Alternatively it is possible that pupils are not able to interpret data as generalized information and vice versa, ie are unable to 'translate' it into generalized relationships expressed verbally. This may apply even if they have the skills to extract individual data points. In either case it implies that pupils' skills of knowing what it is to *interpret* data or to *generalize* from data require more specific and perhaps explicit development.

Generalization from complex 'multivariate' data

Within the group of questions requiring generalization, the lowest performance is generally found when complex multivariate data are presented, and pupils are invited to decide which variable is functional in altering another. Such a question is exemplified by 'Birds' eggs (2)' (Figure 8.4) which presents pupils with tabular data on egg size, incubation period and clutch size for the eggs of various species and asks pupils to state any pattern. In effect such questions require pupils to perform what has been termed a 'natural experiment', ie the manipulation of given variable-based data (rather than the generation of such data as occurs in a 'planned' experiment – Rowell, 1984). No information is available on the strategies used by pupils to generate their responses.

Questions of this type can produce low scoring even from the most able. Sometimes less than 60 per cent of this group choose the correct option, compared to the 90 per cent plus producing adequate attempts at generalization in more routine circumstances. There is evidence that the difference is composed of able pupils who 'generalize' the data but have difficulty in effectively eliminating irrelevant variables. Usually in questions of this type the item stem states very explicitly that the response expected to the question is the extraction of a relationship between variables. Thus, despite the complexity of the data, middle and low performance groups include quite high proportions of pupils who attempt to generalize.

However, the strategies required to extract specific *functional* variables (and eliminate others) pose particular problems for these groups and decreasing proportions of those who generalized did so correctly. These points can be illustrated by the responses to 'Birds' eggs (2)' shown in Table 8.10.

Figure 8.4 *'Birds' eggs (2)'*

The table shows the eggs of 5 birds (drawn half-size). Underneath you can see the number of eggs usually laid at one time. The average number of days to hatch the eggs is also given.

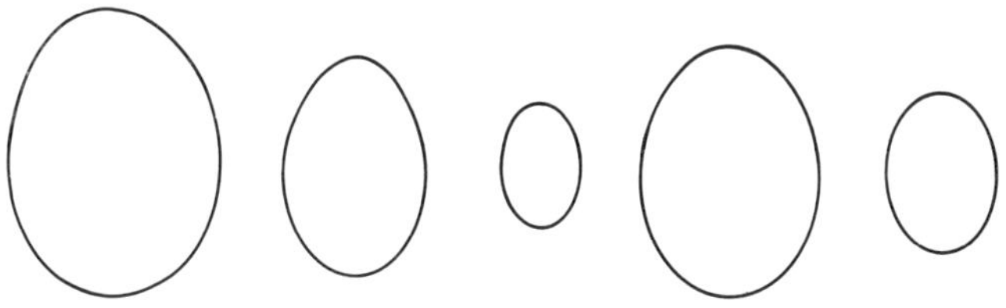

	Golden eagle	Crow	Robin	Wild duck	Blackbird
Average number of eggs	2	5	6	12	4
Average number of days to hatch	40	19	13	30	14

(a) What pattern do you notice in these results?

....

....

....

....

....

(b) This is the size of a magpie's egg (on the same scale).

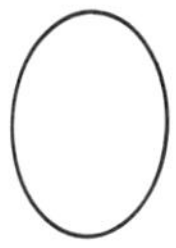

On average it lays 7 eggs.

About how long are magpies' eggs likely to take to hatch?

So far as the most able are concerned, performance on these questions appears to be sensitive to the complexity of the manipulations involved. This contrasts with their *generalized* skills in more routine questions. The presentation of data in an ordered form generates higher levels of performance than the presentation of essentially random data points. The questions 'Heart beat survey (2)' and 'Lichen (3)', which have a similar logical structure, present data as shown in Figure 8.5.

Table 8.10 *Responses to the question 'Birds' eggs (2)' by performance group*
(Percentage of performance group giving indicated response)

	Attempt at generalization			Only cites specific cases	Other	No response
Performance group	Total	Correct	Failure to exclude clutch size			
Most able	94	78	16	0	3	3
Above average	82	69	13	3	9	7
Average	89	59	30	0	7	4
Below average	56	31	25	3	19	22
Least able	30	11	19	2	31	37

Figure 8.5 *'Heart beat survey (2)'*

Name	Age (years)	Heart beat rate (beats per minute)
Tim	4	100
Jane	4	110
Carol	5	100
Simon	5	95
John	5	92
Greg	7	84
Susan	7	88
Stephen	9	80
Ann	9	85

'Lichen (3)':

Diameter of lichen patch	Date on the gravestone	Direction in which the stone faced
cm		
2.0	1958	South
6.0	1918	South
1.0	1968	South
1.0	1958	South
2.0	1958	South
2.0	1938	North
0.5	1968	North
7.0	1908	South

Pupils are asked to decide whether age and/or sex in the one case, and age and/or direction in the other affect the dependent variable. The dependent variables are identified as 'heart beat rate' and 'diameter of lichen patch' respectively. Seventy-nine per cent of the most able identified both variables as functional in the first case, despite the greater obscurity of the variable 'sex', and only 56 per cent in the second.

Many pupils who appear readily able to generalize a two-variable relationship from presented data do not undertake the extra step of manipulating data based on one dependent and two independent variables in order to extract variable relationships. This may be due to innate cognitive difficulty. However, it is doubtful whether such variable handling tasks receive attention in normal science education practice.

Prediction of new data

Questions involving prediction are generally more diverse than those involving generalization or description. It is frequently possible to ask pupils to make predictions on data about individual cases (using their own knowledge of the phenomena involved), where generalization from the data presented would not be reasonable. One question of interest here is whether pupils must explicitly generalize from data before they can make predictions. It is possible that pupils can *appreciate* data sufficiently to use and make predictions without being able to express generalizations.

Performance on predictive questions was generally poorest where relatively complex manipulation of numerical values was involved. The presence of quantitative relationships was not, however, a sufficient cause of poor performance: simple extrapolation or interpolation of ordered bivariate data was usually undertaken efficiently. However performance was sensitive to complicating elements such as disordered data or problematic qualitative variables. Table 8.11 shows the percentages of pupils choosing the correct options in a number of questions which varied in the ordering of data and in the type of variables present.

In fact, the performance of the least able approached a random distribution in 'Celery'. The impact of disordering data was considerably greater than that of the introduction of a non-numerical variable (in 'Celery'). This may be due to the difficulty of reordering the data, but it could also reflect a failure to perceive the *importance* of sequencing data. Thus the least able often seem rather to focus on the data point which is closest to that present in the question demand and undertake some intuitive adjustment, usually based on a direct proportionality between the quantities involved. In the questions 'Indcomp' and 'Celery', referred to above, the percentages making predictions of new values consistent with (though of course not necessarily *due to*) this strategy are shown in Table 8.12 (p73). This table also shows a quite common pattern for complex questions in which there is limited variation among the four lower ability groups and a contrast between these and the most able.

Again, the increase in complexity which results from introducing multi-variate data, so that pupils must decide on the important variable relationship before making their prediction, appears to have an important role in determining performance even among the most able. In the question 'Pollen (3)' (Driver *et al*, 1984,

Table 8.11 *Performance in three predictive questions involving extrapolation and interpolation*
(Percentage of performance group giving correct response)

Performance group	Question: 'Carspeed' (ordered numerical data)	'Indcomp' (disordered numerical data)	'Celery' (disordered data involving spatial variable)
Most able	90	73	67
Above average	85	45	43
Average	78	33	38
Below average	67	35	32
Least able	55	24	18

Table 8.12 *Choice of options in 'Indcomp' and 'Celery'*
(Percentage of performance group choosing incorrect options consistent with matching single data points)

Performance group	'Indcomp'	'Celery'
Most able	19	14
Above average	38	34
Average	47	41
Below average	32	33
Least able	35	40

p72), where the prediction is based on disordered data for pollen count, temperature and humidity, only 35 per cent of the most able selected the correct option. Here it was necessary *first* to decide which independent variable (temperature or humidity) affected the pollen count and then make the prediction.

The strategies used by pupils in making predictions are revealed only to a limited extent from the final outcome of the process. It is possible that pupils first *generalize* and then undertake deductive reasoning, but this type of 'expert' reconstruction may be misleading. Certain of the questions (question types 7 and 8) ask pupils to make a prediction and then justify the prediction which they have made, and, of these, a proportion have been marked so as to indicate the reasoning patterns pupils suggested. The report of the 1982 survey presented data on the relationship between prediction and justification for the question 'Shadows' (Gott *et al*, 1985, pp80–83). These data were interpreted as showing that pupils who failed to suggest generalizations based on the given data also failed to predict a new instance successfully. Analysis of questions with a slightly different technical structure indicates more of the detail of pupils' approach to this type of task. The relationship between pupils' predictions and their justifications are shown in Table 8.13 for the question 'Planets' (Driver *et al*, 1982, p87) which presents pupils with tabular numerical data on planets' orbital radii and periods. They are asked to predict the period of a new planet of given orbital radius in the first part of the question, and justify their prediction in the second part.

In 'Planets' it is possible to make an accurate prediction (within the terms of the question) by a strategy of locating the two adjacent values of the dependent variable ('bracketing'). It is evident that many students utilize this strategy rather than first generalizing from the data and then undertaking deductive reasoning. This approach is ineffective at dealing with inverse relationships or those which are not linear.

The distribution of the approaches across the performance groups is shown in Table 8.14, from which it seems that the 'bracketing' approach is dominant even among the most able. Among the less able reference to only one other data point becomes relatively more important, but many pupils either give no response or one which cannot be interpreted as a coherent strategy.

These comments illustrate the difficulties which determine the effectiveness of pupils' attempts at predictive tasks. The problem of understanding the question demand at the most general level appears less significant as a source of low scoring among less able pupils here than in questions requiring pupils to generalize data. Pupils commonly know that they are intended to predict a datum or data, usually of the type illustrated in the item stem. Notions of how to undertake this predictive task – the general idea of *using* the data given to generate new data – often appear to be lacking among the least able. The absence of more specific strategies – such as the actual techniques for the extraction of trends from complex data and of the application of these to specific cases – are more frequently the cause

Table 8.13 *Cross-tabulation of responses for the two parts of the question 'Planets'*
(Percentage of pupils giving the combination of prediction and justification indicated)

	Justification given				
Prediction	Generalization with deduction	Based on two adjacent instances	Based on one instance	Other	No response
Correct	8	27	15	25	6
Incorrect	0	0	<1	10	6
No response	0	0	0	0	2

Table 8.14 *Justifications offered for predictions in 'Planets'*
(Percentage of each performance group offering justification indicated)

Performance group	Generalization with deduction	Based on the two adjacent instances	Based on one instance	Other	No response
Most able	26	45	9	18	2
Above average	10	34	22	30	4
Average	4	30	17	37	12
Below average	2	18	14	47	19
Least able	0	9	14	45	31

of failure among able pupils. Even when predictions are correctly made, the above data indicate that in many cases the strategies are either not understood coherently or not general in their applicability.

The precise nature of the question determines the impact of these problems on predictive performance. The data on 'Planets' and 'Shadows' indicate some consistency in the proportions deploying the various strategies. However, even when the manner in which a pupil has generated a particular option appears clear cut the reality *may* be much more complex. The only method of casting light on such questions is through more detailed investigation of individual pupils' responses to questions of similar types, and research into the methods pupils use when attempting questions.

Interpretation of data is perhaps an undervalued component of the competencies involved in school science. Nevertheless the data presented here suggest that the problems and complexities of interpretation are considerable; even able pupils have difficulties with more complex interpretive tasks. Both they and the less able display quite well defined (though different) failures which are probably amenable to improvement. This applies whether interpretation is an entry skill for the main science disciplines or a major skill area which the science curriculum is expected to develop for its own sake.

8.4 Summary

Interpreting presented information is in some respects not easy to define, yet it represents an important area of activity for both the sciences and many other curricular areas.

The attempts to impose an 'expert'-generated structure on the types of questions used to represent **Interpreting presented information** have revealed the diversity both of questions and responses. The account in this chapter has described 'interpretation' questions as requiring the generalization of given data, the prediction of new data and the justification of such predictions. In addition it has been found useful to adopt a variable-based understanding of the information presented. Most of the questions can be understood formally in this way, though any constraint which would be imposed by attempting to confine information to that containing well defined variables would be inappropriate for reasons of validity. In many of the predictive tasks the information presented is less clear-cut than this. In addition, it is necessary to bear in mind that data which to an 'expert' clearly are based around variables may appear to the pupil as organised differently. Nevertheless, the analysis of pupils' responses on this basis does appear to provide a useful route into the characteristics of pupil performance.

Overall, pupils' performance on **Interpreting presented information** occupies an intermediate position in the assessment framework. However, it is markedly lower than **Using graphs, tables and charts**, indicating the greater demands made in interpretation, and the fact that mere technical competence in extracting and inserting data is not sufficient to allow interpretation. Explicit attention to these higher level skills in data handling is clearly required to generate competence. Wide diversity of performance is observed within the classes of questions referred to. Patterns, however, are observable. Some evidence has been presented to suggest that pupils' predictive performance is improved by the requirement to generalize from data or justify their prediction in association with the prediction itself. In addition it seems that this effect is greater among the less able than the more able, possibly through the more explicit cueing the extra demand provides. The corollary of this is that the failure of less able pupils to score may not be in some cases through innate incapacity to 'interpret' the information. Question-writing strategies which produce a greater involvement in the data, and which indicate *broadly* what kind of methods and response are indicated, can allow the deployment of competencies otherwise unused.

More detailed analysis of pupils' responses reveals systematic 'failures' in pupil perceptions. In questions requiring pupils to generalize data the crudest error observed is that of doing something other (and often more complex) than generalizing. This often involves drawing on knowledge of the phenomena involved. Such responses, rare among the most able, were found to predominate in the least able. Of course, it was also among the least able that simple technical errors due to misreading or inability to understand individual data instances were found. Among groups of middle ability the most extreme failures were replaced by response characteristics such as referring to specific instances, not expressing complex generalized trends accurately and fully, and failing to extract simplifying statements of a qualitatively different kind when these were required. Performance was poorest for all groups (and not particularly good for able pupils) on tasks requiring the deployment of variable handling strategies to extract functional variable relationships. Even the most able appeared to be sensitive to such question characteristics as data ordering in these questions.

Predictive tasks were rarely undertaken by a process of generalizing followed by deduction except among the most able. Many pupils merely matched given data to that about which predictions were required. The strategy of bracketing the latter by given data, followed by interpolation, was the commonest observed. The dangers of this approach, compared to one in which generalization from data is followed by deduction, are that it often ignores lack of linearity or inverse relationships in data.

In the light of this analysis, the major requirements placed on pupils by interpretation questions can be summarized as follows:

— to appreciate the question's requirement to develop a response which is consistent with and constrained by the data (particularly among the least able);

— to conceive data in standard forms as generalizable rather than a mere set of specific data points;

— to use deductive reasoning in predictive tasks;

— to deploy variable separating strategies in multivariate data (a problem even for able pupils);

— to master technical details of reading data points in standard representational forms (this, of course, is more clearly the focus of **Using graphs, tables and charts**).

Each of these requirements for competent performance constitutes a component of interpretational skill, absence of which can be connected with poor performance. These components are suitable for further exploration in two directions. Firstly it would be of value to elucidate the circumstances in which they are deployed by the various ability groups, and the complex ways in which they combine to determine performance on specific questions. Secondly it can perhaps be assumed that, even without an understanding of their interaction in specific cases, the development of teaching strategies to improve performance on each component would be likely to improve overall performance on **Interpreting presented information**.

9

Applying science concepts

9.1 Introduction

The place and use of questions requiring knowledge of 'taught science'

The location of questions which explicitly required an understanding of taught science concepts received considerable attention during the setting up of the APU assessment framework. It was finally decided that they should be confined to the subcategory **Applying science concepts**. This subcategory is subdivided into three broad conceptual areas based on the three major disciplines (see Chapter 1, Table 1.2). Rather than asking pupils to state or explain scientific ideas the questions used here require the application of such ideas in situations representative both of normal school science and of more wide ranging settings.

The concepts of science, insofar as they can be understood in isolation from concrete examples of their usage, are represented by a set of 'concept statements' of varying levels of complexity at the various ages. That at age 15 is the most complex, and is intended to represent the field of knowledge to which a child following a basic science course might have been exposed by the end of the third year of secondary school (see Appendix 7). Any such list will of course be a compromise. It cannot be guaranteed that every child will have been exposed to the ideas required to answer any given question. Moreover for pupils studying the sciences after the third year of secondary school, repeated treatment and elaboration of conceptual areas will be expected to have an effect on understanding. The list of concept statements was validated by extensive consultation with teachers, LEAs and examination boards. The statements were used by question writers as the guidelines and limits to the questions which could be written to represent the subcategory. It should be clear that the set of statements, and the questions written to them, are not intended to represent either a teaching or an examination syllabus.

Since all APU tests are given to a random sample of pupils no attempt can be made to ensure that particular questions are taken only by pupils studying the appropriate science. The obvious problems which this entails are compensated by its potential value in assessing the impact of science curricula on performance, as discussed in Section 9.2.

The structure of the subcategory

The most important divisions within the questions representing the application of science concepts are those between the three science disciplines. Each discipline is represented by a pool of questions. In addition each question is allocated to one of two concept regions. The regions are:

Biology	A	Interaction of living things with their environment
	B	Living things and their life process
Physics	C	Force and field
	D	Transfer of energy
Chemistry	E	The classification and structure of matter
	F	Chemical interactions

There exists a further classification into 'concept pages', generating 28 subdivisions in all (see Appendix 7). Finally questions are allocated to a number of question types as shown in Table 9.1 (p77).

It can be seen from Table 9.1 that the question types are based on the class of activity the pupil is required to undertake, and the mode of response needed. Three such classes of activity have been considered appropriate in the domain of applying science concepts. These are:

— prediction of new data;

— explanation (including forming hypotheses from given or predicted data);

— evaluation of given predictions and explanations.

For purposes of ensuring stability in the pools it is useful to monitor and control the representation of the question types.

The three pools of questions representing **Applying science concepts** are used in various ways. An important use is to generate performance estimates both nationally and for various subpopulations. By this method it is hoped to gain some indication of the relative impact of different variables on competence in these disciplinary areas. Section 9.2 presents such data broken down according to the curricular background of pupils.

The interpretability of such scores is however limited. They give a composite picture of competence on a particular discipline, but tell us little about strengths

Table 9.1 *Distribution of questions on 'Applying science concepts' across question types and discipline*
(Percentage of each question type within indicated discipline)

Question			Discipline		
Type	Demand	Response mode	Biology	Chemistry	Physics
1	Predict and/or explain new or given data	Extended response	9	8	11
2	Predict	Short response	10	8	20
3	Predict and explain	Coded answer and extended response	13	13	14
4	Predict	Coded answer	17	19	16
5	Explain	Extended response	28	20	21
6	Assess explanations of given data	Extended response	4	9	3
7	Explain	Coded answer	11	11	6
8	Assess explanations of given data	Coded answer	9	13	9
		Total number of questions	138	113	160

and weaknesses across the system of concepts. Sections 9.3 and 9.4 therefore look at the type of information on these issues which can be obtained from APU questions and reports. Performance to date on each concept page was reviewed in the 1983 report (Welford *et al*, 1986, Chapter 5). It was shown there that the variations of individual question scores *within* even these quite narrowly defined conceptual areas was much greater than the variation between the mean scores for the areas. This is indicative of the problems of attempting to isolate specific areas as more, or less, well understood. Because of this, attempts have been made to study particular concept areas in depth using batteries of questions. A brief summary of the characteristics of pupils' understanding revealed by this type of activity is given in section 9.3. The reader is also referred there to previous reports at age 15, and to the work of the Children's Learning in Science project (CLIS) for detailed discussion of pupils' performance on, and understanding of, some of the central science concepts.

One conclusion which emerges from this section and the work referred to is that children's usage of concepts is not such that one can assess understanding of a specific concept statement by a single question. Understanding of even quite simple scientific ideas is multifaceted. There is sufficient space in this review to undertake an exploration of such understanding only for one narrowly defined area. This occurs in section 9.4 where pupils' understanding of combustion is analysed in some detail in order to illustrate its complexity. The opportunity is also taken to show in this section how the characteristics of understanding vary across different performance groups. It is intended that the general points made will be illuminative for other conceptual areas.

9.2 Analysis of subcategory scores against curricular and other variables

The gradual development of the bank has meant that the early subcategory scores showed some variation. At that more formative stage quite small numbers of additional questions could alter the balance of the bank, but a considerable degree of stability was obtained after two years and is described in Chapter 4.

The overall subcategory scores, and those for boys and girls are shown in Table 9.2. In order to avoid problems caused by variations in examination entry practice or curriculum patterns across the three participating countries, these and ensuing figures refer to England only (see Chapter 2). In addition, they represent an amalgamation of the data for the years 1982–84. The patterns within each year are qualitatively similar.

Table 9.2 *Performance on 'Applying science concepts', 1982–84, England only*

Subcategory	Number of pupils	% Mean scores		
		Overall	Boys	Girls
Applying biology concepts	3,275	35	34	35
Applying physics concepts	3,505	34	37	30
Applying chemistry concepts	3,274	33	34	32

The differences in overall scores across the three disciplines are slight. Thus the figures suggest that the standards being achieved in each discipline are similar, though the possibility of compensatory effects between such standards and the overall difficulty of the pool of questions cannot be ruled out.

The table also presents, for the sake of completeness, mean scores for boys and girls. These show little variation between the sexes except in the case of Physics, which is a well-established result (Johnson and Murphy, 1986). Further analysis and comment on the relative scores of boys and girls occurs below.

The most obvious method of assessing the effect of science curriculum on performance is to compare the scores of pupils studying and not studying the relevant discipline. There are, however, problems in using the data for these groups. Firstly, many pupils, while not studying an individual discipline, follow a general science or similar course (see Chapter 2). In addition there is a well established relationship between ability (however measured) and take-up of separate science subjects. In general more able pupils study more science and their courses are much more frequently based on the separate disciplines (see Table 2.9). In order to observe the effect of the stated curriculum it is necessary therefore to differentiate between ability groups.

The only available independent measure of ability for pupils taking APU tests is anticipated examination entry. Pupils are classified on this basis according to teachers' judgements at the time of testing. Table 9.3 shows the scores in **Applying science concepts** for pupils in two ability bands: (a) those likely to be entered for six or more O-levels (approximately 30 per cent of all pupils – see Table 2.8), and (b) those likely to be entered for one or two O-levels or four or more CSEs (approximately 38 per cent of all pupils). These correspond with broad spans of the upper and middle ability ranges respectively. Only a very small proportion of the first of these groups, and a somewhat larger proportion of the second, follow general science courses. However, there is evidence that the impact of existing general science courses being taught up to and including 1984

Table 9.3 *Performance according to discipline and anticipated examination entry*
(Percentage mean score on indicated subcategory)

	Examination entry			
	6+ O-levels		1 or 2 O-levels or 4+ CSEs	
Subcategory	course includes relevant discipline	course does not include relevant discipline	course includes relevant discipline	course does not include relevant discipline
Applying biology concepts (n = 2,421)	49	42	32	27
Applying physics concepts (n = 2,622)	52	37	34	25
Applying chemistry concepts (n = 2,453)	52	38	32	24

on **Applying science concepts** is very limited, so allowance is not made for these (see Chapter 13). Table 9.3, therefore, breaks down the score on **Applying science concepts** for each discipline in terms of pupils' ability (as judged by teachers) and whether or not they follow a course in the relevant discipline. For example, the table indicates that pupils likely to be entered for six or more O-levels and studying Physics scored 52 per cent on questions from the Physics pool.

When the disciplines are compared in this way there are again broad similarities: the scores for the most able in all three cases are approximately 50 per cent. This might be considered a relatively low figure for pupils of this type.

A trend is evident for both groups in the relative enhancement in performance among pupils following courses in the relevant different discipline. The differences in scores between able pupils following and not following relevant courses is 15 and 14 percentage points respectively for Physics and Chemistry. This compares with seven percentage points for Biology. The improvement in score for Physics and Chemistry found among those studying these subjects is thus approximately twice that found for Biology. These trends, though quite small, are stable from year to year. This appears to involve two effects. Firstly, in each case performance on biological questions among those not following Biology courses is elevated (relative to the other disciplines). This could perhaps be a reflection of greater familiarity with Biology subject matter and/or the greater accessibility of Biology concepts for the majority of pupils. Secondly, there is a relatively reduced score on **Applying biology concepts** among those following a Biology course, regardless of ability. This may merely reflect the limited value of examination entry as an indicator of ability, ie it may be that the group likely to be entered for six O-levels or more in a course which includes Biology is in fact less able than the equivalent group for other disciplines. In addition it is known that a greater proportion of Biology students study *only* that science, while those studying the other two sciences show a greater tendency to study *both* Physics and Chemistry.

Nevertheless the difference does exist, and it is possible to offer less cautious although speculative explanations. It is of interest to note that Eggleston's work on intellectual transactions in the classroom showed a parallel split between Physics and Chemistry on the one hand, and Biology on the other (Eggleston, 1983). Eggleston's Science Teaching Observation Schedule analysed classroom interactions in terms of both their origin with teacher or pupil, and their intellectual form. By cluster analysis he arrived at three 'types' of classroom practice, which he labelled, 'problem solving'*,

* It should be noted that 'problem solving' as used here is different from the practical problem solving which is discussed elsewhere in APU reports.

'fact acquiring' and 'pupil-centred enquiring'. Physics and Chemistry classes were more likely to be of the first type, Biology classes of the second. APU questions in these three subcategories tend to parallel the more challenging and exploratory activity in the first type of lesson than the second. There is little demand for simple recall of concepts and the questions in general tend not to conform to stereotypes to be found in the classroom. Prediction of new data and explanation of non-standard situations are more common.

There is at least a possibility that the performance data reported above reflect the difference detected by Eggleston and that the most common practice in Physics and Chemistry lessons provides a better preparation for non-standard tasks. Pupils more used to a 'fact-acquiring' Biology approach may be less well prepared to undertake questions such as those found in the APU pools. It must be observed here that section 9.4 indicates a limited transference of conceptual knowledge in a chemical field into a non-standard situation. The success observed here is a *relative* one, and is not intended to suggest that performance in the chemical or physical field on such questions does not indicate problems in transferring understanding and the application of knowledge.

Differences in performance between boys and girls

The scores for boys and girls presented earlier cannot be interpreted without further analysis to allow for the combined effect of curriculum and ability. Since more boys study Physics, for example, it is anticipated that boys will perform better as a group when applying Physics concepts. However those girls who do study Physics are, in general, more able than boys studying Physics. Thus in a typical sample of pupils studying Physics 28 per cent of the boys and 52 per cent of the girls were anticipated to be entered for eight or more O-levels. In Table 9.4 the scores for **Applying science concepts** are broken down by gender, curriculum and ability. The pattern across disciplines here is similar to that discussed above.

The most significant effect observable in Table 9.4 is the fact that the observed differences in performance between boys and girls in Physics are not removed by allowing for curriculum and ability. Girls likely to be entered for six or more O-levels and studying Physics perform substantially less well than the equivalent group of boys. The difference is only a little less than that for the population as a whole, with boys obtaining on average a score one quarter greater than that of the girls. The situation in Biology is in some respects even more noticeable. In the entire population the scores for boys and girls are almost identical. When ability and curriculum are controlled a difference in favour of boys appears for the most able, though this is much reduced for the group of middle ability. Thus the equality of score overall is shown to be a consequence of the fact that more girls study Biology.

Table 9.4 *Performance according to curriculum, anticipated examination entry and sex*

(Percentage mean score in indicated subcategory)

		Examination entry			
		6+ O-levels		1 or 2 O-levels or 4+ CSEs	
Sub-category		course includes relevant discipline	course does not include relevant discipline	course includes relevant discipline	course does not include relevant discipline
Applying biology concepts	Boys	51	44	32	27
	Girls	48	40	31	26
Applying physics concepts	Boys	54	43	35	27
	Girls	43	35	29	23
Applying chemistry concepts	Boys	53	39	32	25
	Girls	50	38	32	22

The observation that able girls perform less well than able boys across all three sciences is disconcerting. The possible effect of the fact that girls generally study fewer sciences than boys must be noted. Unless a view of innate gender-based differences in ability is adopted, it appears that affective factors may have a residual role in depressing girls' science performance. This applies even when positive decisions have been taken by girls to study particular sciences and even within 'traditional' girls' activites. Such affective factors could include the context and emphasis of APU questions as well as the learning experiences provided, and to which girls are directed, within and outside school (Johnson and Murphy, 1986). Still more significant is the opting out of science which follows, as early poor performance reinforces demotivation (see Table 2.13). Together these two effects make this a major educational problem.

9.3 Pupils' understanding of particular conceptual areas

The most obvious method of gaining information on pupils' understanding of specific areas is to subdivide the pools of questions according to the conceptual area involved. Such a breakdown has been presented in one of the research reports (Welford *et al*, 1986, Chapter 5). Its most noticeable characteristic is the uniformity of overall mean scores across concept areas and the diversity of mean scores for individual questions. It is doubtful whether an approach of this kind will provide useful information about pupils' understandings. An alternative approach is to look in detail at pupils' understanding of conceptual areas. This involves using mark schemes which record the categories of responses explicitly, enabling a description of these responses to be undertaken. It is necessary to map a set of questions used for this purpose against the conceptual structure of the field being assessed, and to cover the latter in sufficient detail. As described in Section 9.1 it is unlikely that the pools will contain a sufficient number of questions, or questions of the correct form, to produce such mapping for any given area. All conceivable areas would need to be represented in similar detail in this way within the pools if an imbalance between conceptual areas was not to be produced. It is also improbable that the set of questions drawn randomly from the pools for any given survey will contain all of those available on the particular area.

Despite these problems, it has been considered important to exploit the large body of data available to look at pupils' performance in various concept areas. The following approaches have been used:

- broad surveying of conceptual areas exposing some 'representative' questions from those specifically drawn from the question bank for a survey. This was undertaken in the report of the 1980 survey (Driver *et al*, 1982, Chapter 7);
- the review of all those questions from a single concept area drawn at random for the survey from the bank. This was undertaken in 'The Earth in Space' (Welford *et al*, 1986, Chapter 5) and 'Reproduction' (Gott *et al*, 1985, Chapter 7);
- the use of questions drawn from the bank supported by other similar questions specifically written for this purpose ('probe questions'). This approach was used in the survey of 'Science concepts related to the topic "air" ' (Driver *et al*, 1984, Chapter 8);
- use of a mixture of survey questions and probes but drawing on activities tested by other categories in the science framework. 'Electricity' (Gott *et al*, 1985, Chapter 8) was constructed in this way;
- the use of specially written questions of a different type from those to be found in the bank: the approach used in 'Force' and 'Acceleration, Gravity and Work', (Welford *et al*, 1986, Chapters 6 and 7).

The following section surveys the findings which have emerged from these studies.

Applying biology concepts

Processes involved in maintaining a balance between the component gases in air, and in plant nutrition

The report of the 1980 survey (Driver *et al*, 1982) suggested that about 30 per cent of pupils appreciated that respiration occurs in plants as well as in animals. More than half the pupils tested knew that oxygen is taken in by animals for use in respiration and with the subsequent release of carbon dioxide. Amplification of these findings occurs in the 1981 survey report (Driver *et al*, 1984). About a third of pupils identified that most living things take in oxygen to be used in respiration. A similar proportion stated that carbon dioxide is produced by respiration. About a quarter of the pupils tested stated that in photosynthesis green plants take in carbon dioxide in the presence of light and between a quarter and a third said that oxygen is produced in the process. However, fewer than one in ten of the test sample could successfully marshal the ideas relating to gas exchange in plants and relate them to plant growth.

Pupils' ideas specifically about plant nutrition have been subjected to rigorous investigation by other researchers (Bell and Brook, 1984) to show that between a fifth and a third of 15 year olds used scientifically accepted ideas on plant nutrition in written responses to three APU questions. The ideas about plants manufacturing their food – ie the links between the gaseous exchanges of photosynthesis (and respiration) and the other physical and chemical processes carried out by plants (growth, water uptake etc) – were confused in most pupils' minds.

Interdependence of living things

Early findings were that over half the pupils were able to make simple predictions based on trophic level information contained in a food web, but that fewer than 20 per cent could make predictions when considering more than one effect on potentially stable populations. Senior (1983) reviewed a range of 'interdependence' questions to show that performance in this area is influenced by pupils' familiarity with the organisms named in a food chain or web, and by the complexity of trophic interactions illustrated by a food web. Few pupils have shown a wider appreciation of the effects resulting from ecological disturbance beyond the adjacent element in a food chain. This is presumably because few pupils appeared to realise that a particular organism may be both predator and prey; it may be preyed on by several animals and it may itself eat a variety of organisms.

Reproduction and heredity

Pupils more often correctly identified sexual reproduction when occurring in animals than when occurring in plants. Some popular misconceptions were detected which diverted pupils away from considering points relating to modes of reproduction as they apply to classifying vertebrates. For instance, most pupils decided that the possession of a slimy skin indicated that an animal was a reptile, even when the reproductive differences between amphibians and reptiles were clearly outlined in an animal classification question.

Sexual reproduction was assumed to be a precondition for egg production in poultry – an idea which appeared to override correct notions about reproduction. Such popular ideas are more common among the lower attainers. Only a few (the most able) children correctly identified sexual reproduction as a cause of variation between individuals, although most recognised that natural variation exists among members of the same species. Precise knowledge relating to the hormonal control of ovulation, the release of human egg cells and location of fertilisation, were better understood by girls than by boys.

Applying chemistry concepts

Reactions involving atmospheric oxygen

The findings of the first survey that nearly 70 per cent of pupils knew that oxygen was needed to sustain combustion and that about 50 per cent could identify air and water as necessary for rusting to take place were investigated in greater depth in the 1981 survey report (Driver *et al*, 1984). A larger number of questions was used in that second survey and showed that between 55 per cent and 80 per cent correctly identified or described in general terms the requirement for oxygen to support combustion. Oxidation of metal by atmospheric oxygen as an example of corrosion was similarly described in general terms by between 25 per cent and 60 per cent of pupils. Performance was much lower on other questions which tested pupils' detailed knowledge about combustion reactions in oxygen or air (1–20 per cent), and about conservation of mass (6–38 per cent) during chemical change.

In a study of aspects of secondary students' understanding of elementary ideas in Chemistry based on APU science questions a list has been produced of the various ways in which pupils may not give the anticipated 'correct' answers to questions (Holding, 1985). Pupils tended to confuse elements, compounds and mixtures, and were unable to distinguish between diagrammatic representations of compounds and those representing mixtures or molecular elements. Confusion about the relationship between changes of state and chemical changes also exists widely.

(Section 9.4 of this chapter (p82) looks in more detail at the topic of combustion with particular reference to characteristic responses of pupils of different performance bands.)

Applying physics concepts

Electricity

A recent review of pupil performance on basic circuit skills, simple applications of circuit ideas, predicting current values and understanding of current has produced some important insights into pupil performance (Gott, 1984).

Thus, for instance, a quarter of pupils used the idea of constant current in a familiar example involving two bulbs in a circuit, but fewer than one in ten did so in the less familiar case of five bulbs in series. Some 10 per cent of pupils thought that positive and negative currents follow from the ends of a battery and 'clash' in the bulb. Up to half the 15 year olds tested gave responses based on a using up of 'something' in a circuit. The 'something' was described in terms of power, energy and electricity, but these terms were used with imprecise or 'everyday' meaning.

Some physical properties of air

In questions used in the 1980 survey about a quarter of pupils demonstrated an understanding of air pressure and could explain gas pressure very generally in terms of simple particle theory. More detailed investigations of pupils' understanding of these ideas were presented in the 1981 survey report (Driver *et al*, 1984) and in Brook *et al*, 1983. Fifty per cent of pupils could demonstrate in general terms a knowledge that pressure at a point in a motionless fluid is the same in all directions. About one in ten correctly applied ideas about the movement of air from regions of higher to lower pressure. A similar proportion indicated that the behaviour of gases can be explained if it is assumed that they are composed of discrete particles in constant motion.

The 1982 survey (Gott *et al*, 1985) looked at the use of simple equations of the type $a = b/c$ in formulae for macroscopic pressure and velocity. It would seem that the majority of 15 year olds could apply the equations when the variables involved were velocity, distance and time (with simple numbers presented in a ready-to-use form). However the variables associated with macroscopic pressure (pressure, force and area) presented pupils with many more problems, again despite the limitation to numbers in a simple ready-to-use form. The differences in performance were striking: mean scores of around 30 per cent on questions on pressure compared with 80 per cent for questions on velocity.

The Earth in space

All the questions used to date in the topic 'The Earth in Space' were presented with accompanying performance data in the 1983 survey report (Welford *et al*, 1986). It was concluded that the ideas relating to the Earth's spin, its tilted axis and its orbit around the Sun were not well understood. The low performance on the questions in this topic might indicate a relative lack of exposure of pupils to such ideas. It appears that 15 year olds had difficulty in combining their everyday knowledge correctly with ideas of a spinning Earth also exhibiting translational motion.

Force and acceleration

A comprehensive probe into pupils' performance in this complex area of Physics was presented in the 1983 survey report (Welford *et al*, 1986). Among other findings it was shown that the majority of pupils (70+ per cent) had non-Newtonian ideas about movement and gravity. Basic descriptions of force as a pull or push or as causing change in motion were given by about 40 per cent of 15 year olds although only 25 per cent of pupils' responses avoided confusion with energy, pressure and so on. Fewer than one in five were able to calculate acceleration from data involving mass and force.

Demonstration of ability to recall and state equations defining and relating velocity, acceleration, mass, work and power varied, with between six per cent and 15 per cent satisfying the marking criteria. Performance on questions in this area as a whole was extremely low, with only three or four per cent of pupils being able to define terms such as force, mass and acceleration.

Other concepts in Physics

Questions relating to light, movement and deformation, electrostatics and magnetism and work and energy, have also been included in all the surveys to date, but responses have not been subjected to the detailed examination described thus far. However other researchers have investigated these topics using APU survey questions. The topics studied have included Energy (Brook and Driver, 1984) and Heat (Brook *et al*, 1983). The conclusions reached suggest that few pupils tended to use the idea of energy conservation spontaneously, preferring to focus on tangible aspects of systems such as height or speed when constructing their explanations. Pupils do not use a single coherent view of heat. Fewer than one in twenty gave complete responses based on accepted ideas about change in state and few used such ideas when explaining, say, cooling by evaporation.

9.4 A review of pupils' understanding of combustion across different performance groups

This section looks in detail at pupils' performance within the specific conceptual area of combustion. In addition, it surveys understanding of combustion across different performance groups. The performance groups used are obtained from pupils' *total* score on the APU test package they undertook. These scores are then divided to produce five approximately equal groups of pupils. The questions referred to, with the exception of 'Chem-change', are drawn from the report of the 1981 survey (Driver *et al*, 1984, Chapter 8). It is anticipated that the points made on this area will have a more wide-ranging significance for an understanding of the performance of pupils of different ability levels.

Combustion provides an important route into descriptive and theoretical approaches to chemical phenomena. In addition it represents a common chemical transformation, of which most pupils have experience. This being the case it is of interest to establish how pupils classify the phenomenon in theoretical terms. The question 'Chem-change' invited pupils to identify which one of a number of phenomena was an example of a chemical reaction. The choices made are shown in Table 9.5 for each performance group. The phrases used are those present in the question.

It is striking that only among the most able group did the majority identify burning as a chemical reaction, and that the greatest shift in performance is between this group and the next. Further analysis shows that 40 per cent of those studying 6+ O-levels and whose course included Chemistry chose something other than combustion to illustrate a chemical reaction. It is not intended to argue that mere formal definitions of 'chemical' and 'physical' changes are central to understanding. The distinction has received less attention in chemistry teaching during recent years. Nevertheless it seems to be essential that pupils have some conception of chemical change, which they can apply to simple cases such as those suggested here. Without it the more sophisticated characteristics of chemical changes may prove inaccessible. This appears to be supported in the following data.

The percentage of pupils identifying that air or oxygen is required for combustion to occur has been assessed in a number of questions and is generally quite high. The percentages of pupils identifying *removal* of air or oxygen as reasons for combustion *ceasing* in the questions 'Fire', and 'Cotton wool candle' are shown in Table 9.6 (p83)

Table 9.5 *Identification of a chemical reaction in 'Chem-change'**
(Percentage of pupils in each performance group giving indicated response)

	Performance group				
Change classified as 'chemical'	most able	above average	average	below average	least able
Paper burns to ash and smoke	**70**	**29**	**19**	**18**	**9**
Ice melts in a fizzy drink	12	29	21	21	20
Salt dissolves in hot soup	8	21	26	24	27
Water condenses on a cold window	8	17	21	21	25
Water evaporates as washing dries	2	3	7	10	6
No response	1	0	7	5	11

*Total number of pupils: 577.

Table 9.6 *Explanation of combustion ceasing in terms of the removal of air or oxygen*
(Percentage of pupils in each performance group giving this response)

Question	Number of pupils	Performance group: most able	above average	average	below average	least able
'Fire'	745	92	68	52	41	27
'Cotton wool candle'	769	93	92	91	81	54

In 'Fire' the presence of an alternative expressed in 'everyday' language ('damping down') proved the greatest distractor for the least able. This resulted in reduced performance despite the fixed response format. The latter generally produces an improvement in performance for less able pupils, but it appears that an 'everyday' distractor masked real knowledge in this case.

Combustion begins to gain a deeper theoretical significance when understood quantitatively. 'Iron wool' asked pupils to predict and explain the change in mass on burning a sample of iron wool. Table 9.7 shows the responses.

The polarization between the most able group and the remainder is again very noticeable. A large majority within the latter groups predicted a mass decrease and this represents a very common view even among the most able. It does *not* appear that the suggestion of a mass decrease stemmed from ideas of loss of iron oxide during burning. Thus, while the majority of children are familiar with the idea of air or oxygen as a requirement for combustion, it seems that far fewer pupils see the involvement of the oxygen in a specifically *chemical* way, with new products and quantitative effects. Significant numbers revert to notions of a residual ash, or focus on *physical* changes, in the slightly novel situation of 'Iron wool'. Many pupils appear to see oxygen as *enabling* but not quantitatively *participating in* 'combustion'.

In the question 'Rusty nails', pupils were asked to predict and explain the change in mass which would occur if iron nails were allowed to rust over a period of a year. The results are shown in Table 9.8.

There are clear parallels here with the situation in 'Iron wool', in that the use of oxygen-based ideas in a coherent fashion is achieved by about half the most able group. The proportion is much smaller in other groups. It is noticeable also that the idea of a mass decrease continues to be the most attractive alternative for the remaining groups. In both 'Iron wool' and 'Rusty nails' the majority of pupils appear to consider the formation of a new product not to result in a mass increase. Attention is focused mainly on the mass of the iron with notions of iron being 'burnt off' and so on often suggested. In both cases this appears to be consistent with the observed failure in 'Chem-change' to

Table 9.7 *Prediction and explanation of mass change when iron wool is burned**
(Percentage of pupils in each performance group giving indicated response)

Prediction	Explanation	Performance group: most able	above average	average	below average	least able
Mass increase	'oxygen-based' explanation	46	7	6	2	0
	other explanation	6	10	9	14	14
Mass decrease	physical loss of product	7	5	1	0	1
	other physical loss	25	41	34	25	15
	other explanation	9	23	31	40	32
No change		6	11	11	8	18
No response		1	3	5	12	18

*Total number of pupils: 599.

Table 9.8 *Prediction and explanation of mass change when nails rust**
(Percentage of pupils in each performance group giving indicated response)

Prediction	Explanation	Performance group: most able	above average	average	below average	least able
Mass increase	oxygen based	48	13	2	2	0
	other	12	13	19	21	12
Mass decrease	mechanical loss	10	28	31	26	32
	other	9	15	13	9	19
No change		21	30	32	29	25
No response/uninterpretable		0	1	3	13	12

*Total number of pupils: 797.

distinguish a chemical reaction. It suggests a failure to recognize the formation of chemically distinct substances as opposed to changes of state or other physical change, and appears to support the suggestion made there that a paradigm of chemical (as opposed to physical) change and combination is important for theoretical development.

It can be argued that an appreciation of mass changes during chemical reactions is a corollary of the idea of conservation of mass. The question of the mass of the oxygen used up can otherwise be ignored. Appreciation of the concept of mass seems limited in relation to a substance such as air, which is encountered as an imponderable fluid and, frequently, as an 'infinite' resource. The question 'Phosphorus' looked at pupils' understanding of mass change in a sealed container. It showed only 34 per cent of pupils anticipating conservation of mass with a *limited* volume of air.

Broken down by performance group, the responses were as shown in Table 9.9.

Table 9.9 *Prediction of mass change when phosphorus is burned in a closed container**

(Percentage of pupils in each performance group giving indicated response)

	Performance group				
Prediction	most able	above average	average	below average	least able
Mass increase	6	11	16	14	13
Mass unchanged	64	34	21	17	9
Mass decrease	26	49	52	55	47
Insufficient information	3	4	8	9	16
No response	0	1	2	4	12

*Total number of pupils: 769.

There is again some polarization between the most able and the remainder, with the latter tending to opt for a decrease in mass. Analysis of the responses indicates that changes of state and the using up of atmospheric oxygen interfere with conceptions of mass conservation among the latter groups. Again little attention is paid to the mass of a chemically combined product, so that the 'loss' of air or of phosphorus dominates responses.

The question 'Exhaust' asked pupils to make an estimate of the mass of exhaust gases produced from a car in relation to the mass of petrol used. Formally it required no more complex ideas than those utilised previously. In practice only three per cent of the *most able* group suggested an overall increase in mass, with 55 per cent suggesting that the mass of exhaust gases would equal that of the petrol used and 27 per cent that it would be less. The latter figures are similar for the remaining performance groups. It appears that even the most able did not interpret this situation in terms of chemical combination. This may be due, for example, to the absence of a visible source of extra mass or a tangible manifestation of this particular body of exhaust gases. (Between five and ten per cent of each group indicated that exhaust gases weigh nothing.) In any event conservation reasoning was commonly used to justify the idea that mass will be unchanged. The mass of the air and the chemically altered nature of the products was ignored. Thus in these trying circumstances even the most able either revert to 'commonsense' notions of combustion or do not connect the phenomena involved with combustion.

Overall, in these questions, a bare majority of the least able, and almost all of other ability groups, identified air or oxygen as a participant in combustion. About half of the most able, and much smaller numbers of other groups, elaborated the quantitative implications of this in the routine circumstances of 'Nails' and 'Iron wool'. A very small proportion of all groups was able to apply it in the unconstrained and novel situation of 'Exhaust'.

9.5 Summary

A large number of questions have been used as the basis of discussion of pupils' performance on **Applying science concepts**. Questions have been grouped by question type and concept to be applied. Performance within any of the groupings used shows much internal variation.

In considering performance across the whole category it has been noted that the overall estimated mean scores for **Applying biology, physics and chemistry** concepts are very similar. This suggests that the pools of questions for each of the subcategories carry a similar mix of 'difficulty' of question. Overall boys perform better than girls when applying physical science concepts with a slight difference favouring girls in areas of biological science. The effect of studying the discipline in question is to enhance performance, as assessed through mean score. However since both the gender and the ability of pupils confound interpretation of performance (those studying the three main sciences in any combination tending to be of higher ability; girls studying Physics tend to be from the higher ability groups) such comparisons need extensive qualification.

The established gender differences in favour of boys, especially in Physics, are again noted, and shown to persist despite comparing boys and girls of similar curriculum background and ability.

The types of classroom interaction encountered by pupils studying Biology have been suggested as being different from those engaged in by those studying physical science (Eggleston, 1983). APU figures show that for given ability groups the enhancement of performance among those studying Biology on **Applying of biology concepts** is less than the corresponding enhancement of performance within the physical sciences. This is particularly the case when comparing the highest achieving pupils of both sexes. It is suggested that there may be a connection between the styles of certain classroom interactions and the way in which APU questions in this category expect pupils to apply their scientific knowledge.

A summary of the concept areas reported by APU (and others) in previous reports is presented together with a more detailed analysis of the area of combustion. These attempts to elucidate children's conceptual understanding provide insights which are an aid to practice, in terms, for example, of sensitizing teachers to generalized problems in pupils' understandings. However, the difficulties of such activity are considerable, and this difficulty is in some respects informative.

Any attempt to investigate children's conceptual understanding is committed to a degree of generalization. At the minimum level this involves assimilating the diversity of pupils' responses to a scheme of response categories. The interpretative character of this activity, and the assumptions involved in mapping responses to such 'expert'-generated categories should not be underestimated. The most obvious conclusion which can be drawn from APU data is the difficulty of generalizing, in the sense of stating that individual pupils have (or have not) certain concepts.

The first aspect of this difficulty is the existence of various ways in whch conceptual understandings may be said to exist. For example, pupils who know that oxygen is required for combustion may understand this in a manner very different from that involved in chemical combination. The evidence from APU data is that this is the situation for many pupils. Can a pupil be said to understand oxygen-based combustion when evidently unaware of the formation of chemically new substances or of the relevant mass changes involved? To what extent is the elaboration of this set of ideas (particularly in its quantitative aspects) as a theoretical whole required before the pupil may be said to 'understand' oxygen-based combustion. Parallel if more complex states of affairs can be found in pupils' understanding of photosynthesis, ecological interdependence and the relationship between pressure and fluid movement. Large proportions of pupils can be found who understand the basic element or elements within the conceptual structure. The requirement to develop these elements or to marshal them into a whole, produces evidence that they are 'understood' in a piecemeal sense, and often with a different theoretical meaning from that of 'experts'.

It is often only among the most able that anything resembling a developed as opposed to a routine and piecemeal deployment of the disciplinary concepts of science is displayed. Again, in the context of combustion only half of the most able pupils (and very small percentages of other groups) appear able confidently to function with the elaboration into quantitative effects. The second performance group is in this respect closer to the least able than to the most able. Though other conceptual areas have not been studied from this perspective, this type of relative ability 'profile' seems likely to be not uncommon, even though the absolute percentages vary from area to area.

The second problem in generalization on the model of 'concept possession' is illustrated by the situation in 'Exhaust', but other instances could be used. When the phenomena involved are obscure (in various possible respects) understanding can appear to collapse. This characteristic is found even among the most able. Two interpretations of this are possible. Understanding of the idea of combustion may function only within a limited range of phenomena. Alternatively, of course, the operation of a petrol engine may not be perceived as an instance of combustion. Whichever of these is in fact the case, they again highlight a problem which runs generally through APU findings in the conceptual domain, that the transference of novel 'understanding' beyond the immediate environment in which it is produced is very limited. This can only be observed as a problem for the most able, where there is understanding to be transferred.

These findings indicate that it is possible to identify the broad characteristics of children's collective responses in an informative way, but that the notion of 'conceptual understanding' should always be understood critically. They reinforce indications from the profile of subcategory scores across the framework that understanding of the concepts of taught science represents the area of lowest competence and presents the most striking problem within that framework. This last point stems from the fact that, while the 'procedural' subcategories have in general received little explicit attention in science education practice and are open to pedagogic intervention, the conceptual area has been the major focus of attention for many years. A further problem is evident in relation to individual assessment. The specificity of pupils' responses indicates that any instrument is likely to have only limited reliability unless it samples a wide range of phenomena and assessment situations, whether in an APU context or elsewhere.

At the broadest level the implication for practice may be that diversity of illustration and application is of greater importance than is often suggested. This would involve a rejection of the idea that conceptual understanding at a generalized level emerges out of a brief study, however clearly or imaginatively presented, of a limited number of phenomena. It may imply that a reduced conceptual burden is required. This, however, would be consistent with other shifts currently proposed in science education practice, and with the notion that 'procedural' skills might be explicitly developed within new teaching strategies treating a broader range of phenomena. The promotion of more deep-rooted conceptual understanding (albeit of a more limited range of conceptual areas) and of 'procedural' skills might be undertaken together.

10

Planning investigations

10.1 Introduction

Planning of investigations is mainly concerned with planning experiments in the abstract and has undergone significant shifts in internal structure. Originally three subcategories were defined:

assessing testable statements
assessing experimental procedures
devising and describing investigations

The second, and part of the first, of these were subsequently designated as **Planning parts of investigations**, and the last as **Planning entire investigations**. This is the current state of division in the category. A more detailed account of these developments is given in the age 13 review report (Schofield *et al*, 1988).

Planning entire investigations

This subcategory inevitably raises the question of our understanding of 'scientific investigation', more particularly in the context of the school curriculum. This question is of significance also to **Planning parts of investigations**. It involves large scientific and philosophical questions, so that some ground-clearing or simplifying strategies are evidently required. In the approach which has been developed within the APU framework the burden of the activity is borne by tasks having the following characteristics:

— they are concerned particularly with establishing the relationship between variables, thus implying an analytical, variable-based view of phenomena;

— the focus of problems is on the isolation of dependent and independent and control variables and on technical aspects of variable 'operationalizing' (this does not however mean that theoretical ideas about the phenomena involved are irrelevant);

— the role of investigations in theory elaboration or construction is not addressed. The tasks are thus not presented with any reference to this, though a short *practical* rationale of the prescribed investigation is often used.

Tasks of this type also form the basis of the questions set on **Planning parts of investigations**.

The approach to pupils' responses also required methodological decisions.

Pupils' accounts were analysed by the use of checklists specific to each question, compiled after substantial trialling. In constructing these checklists, as in developing questions the approach used was derived from that adopted in the original conceptualization of scientific investigation. Pupils' responses to questions inevitably exhibited substantial internal complexity, therefore a single list of response types was not adequate. This was not necessarily a problem to be overcome, but it constituted a significant source of information for a more informative account of performance. Such an approach, as well as practical considerations of demands on pupil and marker time, made the use of a bank of items in the construction of a subcategory score impracticable. For these reasons questions in this subcategory were approached individually. Pupils' written accounts of whole investigations were not presented as varying along some single assessment dimension or represented by a single test score.

The position of **Planning entire investigations** in the framework has been somewhat overshadowed by that of **Performing investigations**. Pupils' plans should involve classes of activity which are similar to those they actually undertake, so that the methodological problems of analysing responses in the two tasks also have strong similarities. Questions of the validity of such analyses as measuring performance in *generalized* (as opposed to question-specific) scientific investigatory activity are also similar, as are practical limitations on operating within an item bank method. The report of the 1980 survey presented a few written tasks individually. Later reports have generally presented them as an adjunct to **Performing investigations**. This review follows a similar general approach, and comments and performance data on **Planning entire investigations** are presented in Chapter 11.

Planning parts of investigations

For the reasons previously referred to, and for other more positive reasons discussed below, it was decided to assess aspects of experimental planning also with questions of a more analytical type. The questions could be generated and used in greater numbers than those in **Planning entire investigations** and are now located in the single subcategory **Planning parts of investigations**. By focusing on specific aspects of experimental planning these questions provided information on such aspects in isolation, while allowing sampling of a range of contexts and contents. Moreover, they lent themselves to fixed and short response formats. From this more extensive assessment mode it was possible to construct subcategory scores. A second set of issues concerning validity appear at this point. They complement those of the holistic situation in **Planning entire investigations**, with its smaller number of questions and specific question reporting. How were the varying demands made by the more analytical questions to be balanced in the question pool? To what extent does the complex linguistic setting required in such questions modify the (anticipated) 'planning' demand?

In practice the response to both questions has been pragmatic. That of Question Type representation involves decisions concerning the relative importance of different aspects of **Planning investigations** which are difficult, perhaps impossible in principle, to justify. In practice the main components that were identified (see below) have been given roughly equivalent representation, with a few judged to be of less significance. The linguistic problem has been treated on a question-by-question basis. It has been found that questions in this pool do provide useful information on pupils' approach to planning, and this supports the idea that the domain score is informative, if not an absolute measure, in this respect.

Originally two 'analytical' subcategories were defined: **Assessing testable statements** and **Assessing experimental procedures**. Questions in the former focused on the shift from all propositions to propositions which explicitly allowed 'scientific' testing. It was decided to assess pupils' appreciation of this in 'critical' and 'creative' modes. Two sets of questions were devised. The first required pupils to decide on the testability of given statements, and the second the conversion of untestable statements to possible testable 'equivalents'. The subcategory was found to present difficulties. This was partly because problems over pupils' interpretations led to doubts about the testability of statements. In addition, limitations on formulating non-testable statements led to a tendency to linguistic repetition. A subcategory score was produced for **Assessing testable statements** in the 1980 survey, but a performance estimate has not subsequently been generated.

Assessing experimental procedures has been developed as the major location of the more analytical approach to experimental planning, incorporating parts of both the original subcategories and renamed **Planning parts of investigations**. To this end the bank of items was reconstructed before the 1983 survey, with many questions rewritten or replaced. The new bank was assimilated to an analytical structure based on a view of the major logical components of experimental planning. This structure draws on the approach referred to in the previous section and is concerned with the identification, handling and operationalization of dependent, independent and control variables, together with other generalized aspects of experimental activity such as sampling and averaging.

It was not possible to devise questions which focused on these in a wholly analytical manner without generating an idiosyncratic body of items. Question Types were therefore developed which seemed sensibly to assess elements in the planning of investigations, rather than being matched to an exhaustive set of logically determined formats. On this basis it was found that sets of questions having the following anticipated emphases could be developed:

- identify control variables only;
- identify control and independent variables;
- identify or evaluate operationalization of control and independent variables;
- identify dependent/independent variable combinations;
- identify operationalization of dependent variables;
- identify or evaluate methods to ensure validity of operationalization;
- indicate methods for the utilization of data.

Each of these is referred to as a Question Type. In each case questions were generated which covered a range of demands on pupils: they were asked to select and generate responses and evaluate or criticise statements within the areas referred to. Sampling of diverse contexts and phenomena was also undertaken. Each question type also exists in all three response modes (fixed response, short answer, extended response). Examples of questions of each type have been published (Welford *et al*, 1986, Appendix 3).

The extent to which abstract analyses of Question Type and question demand in **Planning parts of investigations** have a role in identifying systematic, generalized aspects of pupil performance and of pupils' perception of tasks is of course a fundamental and difficult research problem. Structured reporting of pupil responses, though potentially useful in the analysis of 'planning' abilities, has been rendered difficult by the diversity of pupil responses. The more tightly defined the framework within which questions are handled the more interpretable, it is anticipated, will be individual and collective scores on any element within that framework. However, the extent of this interpretability is limited, if other aspects of questions, not directly connected with their 'planning' content, have an important role in determining performance. These characteristics may not even be systematically identifiable across questions. Moreover, the greater problems of communication involved in describing pupils' performance by an increased number of scores must also be borne in mind.

In sum, the development of more sophisticated structures for the analysis of the subcategory generates twin questions: on the one hand, the communicability/practicality of the structures, and on the other, their value as observed through the empirically-testable relationship with pupils' responses. The major function of the breakdown has been, so far, to establish a structure within which the subcategory could be understood and to facilitate greater control of the bank of questions by which it is represented. Thus the numbers of each class of question can be controlled across the bank for purposes of comparability. As stated earlier, such decisions can be made on a pragmatic basis, without implying that these are judgements of absolute significance.

The bank of questions covering **Planning parts of investigations** has reached a stable form based on the structure described above with the representation of the Question Types varying between 27 per cent (Question Type 4) and eight per cent (Question Type 2).

10.2 Pupils' performance in 'Planning parts of investigations'

This account is based on the data available from national testing on the 100 questions for which data are currently available. Those for 1983 and 1984 have been amalgamated, since variations in internal structure of the performance data between the two years are slight and well within errors attributable to sampling of questions and pupils.

Overall performance in **Planning parts of investigations** is of course indicated by the score on a random sample of questions drawn from the relevant question bank. The overall mean score for the 100 questions used is 42 per cent. This figure is discussed and further analysed in Chapter 4. In the previous section what might be called a logical 'domain structure' was outlined. Part of the discussion presented here will focus on pupils' performances across the elements within this structure, and the extent to which they can be usefully distinguished. However it is clearly possible to impose numerous other structures across this, or any other, pool of questions, most of which have no necessary connection with **Planning parts of investigations**. They may, however, have an important connection with performance, and they may also be confounded with the domain structure as operationalized in the pool of questions. In the following discussion only one such structure will be identified: response mode. Three broad response modes are defined here:

fixed response		
short response	–	a word or phrase
extended response (prose)	–	at least one coherent sentence

For reasons which will be referred to below the former two are amalgamated, and their distribution across Question Type for those questions used is shown in Table 10.1.

Table 10.1 *Relationship between Question Type and response mode*

Question type	Number of questions used: Fixed response or short answer	Prose
1	19	5
2	15	0
3	9	5
4	10	0
5	4	10
6	1	16
7	0	8
Total	58	42

The confounding between the two classifications is clear. In this section the information obtained from breaking down the overall domain score is discussed, particularly in the light of this situation. Variations in performance by gender and across 'ability' groups, previously defined are also discussed. Some attention is also given to the sources of pupils' failure to gain marks in the various types of question.

Variations in the overall pupil performance across Question Type

The existence of Question Types which focus on different aspects of experimental planning invites an attempt to give a more structured account of performance. Data on this were presented in the previous report for the 1983 survey (Welford *et al*, 1986, Chapter 4). The 1984 data confirm this general pattern, and the overall variations in mean score are presented in Table 10.2.

Table 10.2 *Comparison of performance on the various Question Types*

Question type	Percentage scores: All	Boys	Girls	Number of questions	Score range
1	54	54	55	24	7–82
2	58	58	57	15	28–70
3	45	45	46	14	6–69
4	49	50	48	10	24–74
5	31	31	31	14	2–57
6	20	20	21	17	2–51
7	20	19	21	6	9–34

It can be seen in this table that there are no significant differences between the scores for boys and girls. This is usually the case throughout this subcategory, and in ensuing tables the separate scores will not normally be presented.

The most significant point for an approach to performance through the domain structure used here is the range of mean scores observed within each Question Type. It is evident that the determinants of performance for any question include factors other than those defined by Question Type. One such factor, referred to in the previous section, is response mode. The impact of this, and its confounding with Question Type for **Planning parts of investigations**, is discussed below.

The variations in mean score referred to indicate that the 'domains' defined by the Question Types would require large samples of questions to generate reliable performance estimates. One 'logical' method of amalgamating the question types is to group together questions involving responses concerning the operational details of experimental activity (to include descriptions of how measurements are taken or of sequences of measurements or variable values) on the one hand, and questions requiring responses only in terms of variables in the abstract (ie as characteristics of the situation). This distinction can be mapped against that between Question Types 3, 5, 6 and 7 and 1, 2 and 4 respectively. If the questions are grouped in this way the performance data are as shown in Table 10.3, and appear to indicate superior performance on the questions requiring more abstract responses.

Table 10.3 *Comparison of performance on 'variable handling' and 'operational' questions*

Emphasis of question	Percentage mean score	Number of questions	Score range
Variable handling	54	49	7–82
Operational details	30	51	2–69

The idea that pupils' performance on generalized variable handling is better than on that involving the detailed planning and operationalization of experimental activity has also received some support in the context of **Performance of investigations**. Moreover the greater specificity of operational details might suggest that such questions will be more sensitive to specific gaps in pupils' practical experience. It can be hypothesized that the greater detail involved in mentally modelling practical, manipulative activity compared to the essentially logical demands of abstract variable handling might explain the difference. The methodological problem of response mode must, however, be borne in mind here. So too must the tendency for pupils, when giving extended responses, to assume details of the activity. This is discussed further in Chapter 11, section 11.2.

Response mode and its relationship with Question Type

The responses required are classified into three types as discussed earlier: prose, short answer, and fixed response. Mean scores on questions of these forms are shown in Table 10.4.

The relationships between response mode and mean score is clearly very great, though whether this is due to linguistic demand alone is not known. Pupil inclination or some other factor may also have a role. The similarity between the short response and fixed response means seems to preclude the use of guessing as an

Table 10.4 *Comparison of performance on questions with different response mode*

Response mode	Percentage mean score	Number of questions	Score range
Fixed response	56	38	24–82
Short answer	55	20	28–80
Prose	23	42	2–62

explanation of the difference. These two modes will be amalgamated in the discussion below. Reasons for the confounding previously referred to are evident: it is difficult to focus on practical details of experiments (eg as in Question Type 5) without requiring some extended form of response. Fixed response questions in this area might merely shift the linguistic burden into reading the alternative responses. Questions on the identification of variables of given status generally involve short or objective responses. Again, although it is possible to produce questions which require an explanation of why a particular approach is used, and inviting an extended answer couched in terms of abstract variables, it is not possible to ensure that the child will address the problem in this way. Such questions often generate answers to the substantive problem based on knowledge and speculation. There are questions of this kind on the bank (notably in Question Type 1) but it is not considered appropriate to increase their proportion.

It is evident that the representation of these response modes across Question Types precludes any realistic attempt to separate the impact of response mode and question type. This problem can be overcome to some extent by adopting the broader classification into variable handling and operational questions referred to previously. The results of this are shown in Table 10.5.

Table 10.5 *Performance broken down by focus of questions and response mode*

Focus of question	Response mode	Percentage mean score	Number of questions
Operational details	Fixed response or short answer	54	14
	Prose	21	37
Variable handling	Fixed response or short answer	56	44
	Prose	39	5

It could be argued that, for questions requiring a prose response, the variation across operational details and variable handling remains, though clearly not for fixed response and short answer questions. However the small number of questions focusing on variable handling, and their similarity (they are all of Question Type 1) makes this conclusion speculative.

Question demand, and its interaction with response mode

By 'demand', in this context, is meant the broad requirement made on the pupil either to make some positive suggestion as to how the experiment ought to be carried out or its results used (S), explain some facet of the experiment which has been given (E), or criticise (and perhaps suggest improvements to) an experiment (C). It has been suggested that this characteristic of questions may be associated with differences in performance and, in particular, that demands for criticism of given experiments would depress performance. This might be expected because of the apparent need to compare a given experimental activity with some internally generated version of it. The questions in this pool have been allocated to these demands and the mean scores are shown in Table 10.6. Because of the possibility of interaction with response mode the scores for prose responses only are included.

Table 10.6 *Comparison of performance on questions with differing demand (prose responses only)*

Question demand	Percentage mean scores	Number of questions	Score range
Explain	28	8	2–51
Suggest	25	17	2–46
Criticise	18	17	5–62

The table indicates that these differences in question demand show only a limited match with systematic differences in pupil performance. Nevertheless the requirement to criticise given procedures does appear to reduce performance slightly.

Planning parts of investigations (as represented by this pool of questions) clearly generated a diverse body of responses. There is no evidence that trends in these systematically connect with those elements of **Planning of investigations** which have been logically identified. Indeed, the sheer diversity of response gives little encouragement to the notion of readily definable skills in the context of **Planning parts of investigations**. They suggest also that reliable domain-referenced estimates at the individual level would require very large question pools and tests.

Variations in the structure of performance across performance groups

For the purposes of this report pupils undertaking tests have been allocated to performance groups based on their total scores in APU test packages. Five performance groups, approximately equal in size, were produced for each test package. These groups correlate quite well, though not exactly, with the relative rating of pupils as made by teachers in terms of examination entry.

The extent to which variations in performance were uniform across the identified aspects of the bank was the main interest of this exercise. The 'profile' of performance across question type is similar for each of the performance groups though the polarizations in performance are slightly more marked for the lower performance groups. This effect can be seen most clearly if the Question Types are again amalgamated into the broader classes involving respectively 'abstract' variable handling and operational details within investigations. Performance data on this basis are shown in Table 10.7.

Table 10.7 *Performance of pupils of different abilities on 'variable handling' and 'operational details' questions*

(Percentage mean score for performance group indicated)

Performance group	Variable handling	Operational details	Ratio $\frac{VH}{OD}$
Most able	75	51	1.5
Above average	66	37	1.8
Average	56	29	1.93
Below average	46	21	2.19
Least able	28	12	2.3

It can be seen that performance in variable handling represents, *in relative terms*, a successful area for below average and least able pupils. The problems of visualizing and interpreting the complexities of experimental activity, as opposed to the relatively clear cut issues of variable-based analyses of phenomena, can be offered in explanation of this. In addition the need, in many cases, to give precise and relatively detailed answers to gain any marks can be contrasted with the 'variable handling' questions where identification of a characteristic of a situation will allow some scoring.

Variations in performance according to gender

The overall mean score for **Planning parts of investigations** is almost identical for boys and girls. This is observed also in the mean scores across the different Question Types and that across different response modes. The latter finding contradicts claims that girls' performance on fixed response test items is depressed. This difference in finding could, however, be due to the less stressful environment of APU testing, and there are other possibilities (Murphy, 1982).

The tasks used in **Planning parts of investigations** do not require knowledge of taught science concepts for their completion. Observed gender differences in performance are frequently related to differences in science curricula. Thus it is perhaps to be expected that such differences will be minimized in this context.

10.3 Some generalized errors within pupils' responses

It is a truism that errors are more informative than 'correct' answers in identifying pupils' approaches to questions, since in the latter 'standard' approaches on the part of pupils tend to be assumed. The latter assumption may, of course, be unwarranted in many cases. However a different approach would be needed to detect this. The following section looks briefly at the general manner in which pupils fail to obtain marks, according to Question Type.

Question Type 1: **identify control variables**. Performance on these questions is generally good. A survey of pupils' responses shows that most gain one or two marks when asked to suggest aspects of an experiment to be controlled. However most questions ask for at least three suggestions, while others ask for a discrimination between variables requiring control (contingent variables) and those which are without significance (irrelevant variables). A typical mark distribution is given in Table 10.8, for the question 'Colours' which requires pupils to offer three variables to be controlled. In this case the investigation concerns the visibility of coloured objects, and pupils are asked to suggest factors about the objects which should be controlled.

Table 10.8 *Mark distribution for the question 'Colours'*
(Percentage of performance group gaining indicated mark*)

Performance group	Mark* 0	1	2	3
Most able	0	9	50	41
Above average	4	24	52	20
Average	5	26	40	29
Below average	12	23	50	15
Least able	39	20	36	4
Overall	12	20	46	22

* Mark: equivalent to number of correct variables offered.

Here, even in the lowest performance group over 60 per cent of the population can suggest at least one variable requiring control, and 40 per cent can suggest two. This figure is not exceptionally high. In the question 'Fatty', where pupils are asked to suggest characteristics to be controlled for groups testing different slimming breads, over 90 per cent of the whole population (and 75 per cent of the lowest performers) can suggest at least one variable to be controlled. As the number of variables asked for is increased, it seems likely that the child must possess or construct an increasingly sophisticated conceptual model of the situation. Pupils' broad understanding of what it is to control a variable (even at generally low performance levels) is indicated by gaining one or more marks, and high scoring may well indicate more adequate conceptual models. This interpretation ties in well with the situation in regard to control of variables found in **Performance of investigations**.

Questions involving the distinction between contingent and irrelevant variables (as defined above) are more uniformly answered across performance groups than those requiring the identification of control variables only. Pupils appear to have relatively little difficulty making the *logical* distinction between the two classes of variable. Thus in the question 'Sugar' where the problem is based on factors affecting the rate of solution of sugar in water almost all pupils identify 'the cost of the sugar' as irrelevant. However variables requiring, again, more sophisticated conceptualization (in this case 'amount of stirring') cause many more problems and there is some indication that such confusion can cause failures in variable control strategies.

Question Type 2: **identify control and independent variables**. The additional problems involved in answering these questions are those of *discrimination* between control and independent variables. Thus the majority of pupils can, as in the previous case, identify some variables requiring to be controlled. However many will then go on to transpose control and independent variables, or identify as a second 'independent variable' one which requires control. The extent to which this represents a genuine logical confusion can be called into question even in the latter case. More frequently it represents a failure of shared meanings with the question writer/marker. Thus a child offering an independent variable as a variable to be held constant may mean that this variable should be held constant within a trial. In other cases the child may offer a term describing an entity in the task as requiring to be altered, but not intend it to be understood as a 'variable'. In the question 'Sprouts' pupils are asked to suggest control and independent variables for an experiment to observe the effect of the mass of sprouts on rate of cooking. It is fairly clear, in many cases, that children suggesting that the 'amount of water' be altered (with or without the mass of the sprout) intend this to mean that fresh water should be used each time. The word 'amount' is not intended to refer to a quantitative variable at all but has a meaning closer to that of 'sample'.

Setting up questions so that they focus attention exclusively on a single set of meanings is probably an impractical task. It may not even be possible, given the richness of meaning and association of all words, including technical terms. This is still more likely to be the case given a situation in which meaning is partly determined by aspects of the question context.

More deep rooted failures in perceptions of experimental activity are indicated when children make positive suggestions that an 'independent variable' be held constant. Thus, in 'Sprouts', one child replied 'She may have to take bits of the sprouts to get them to the same mass'. Various interpretations could be placed on this statement, though it appears to be pointing to a kind of 'compensatory verification' approach, ie suggesting the need to make the sprouts all the same mass and show that they then *do* require the same time. In such cases it appears that the logic of variable manipulation in investigatory situations is perceived differently from the standard approach. The notion of manipulation of independent variables appears to be the cause of such fundamental problems in this group of questions and elsewhere.

(Question Type 4 is discussed before Question Type 3 since it fits with the abstract variable handling activity treated so far.)

Question Type 4: **identify dependent/independent variable combinations**. Questions of this type appear to move one stage further along a spectrum of sensitivity to variations in shared meanings. Each response requires the pupil to decide on the suitability of the given experiment to test a stated hypothesis or answer a stated question. Though approximately three in five judgements made by pupils are correct in most questions, the effect of mark schemes is often to increase the polarization across performance groups. In many questions one or two of the required judgements are usually made correctly by a large majority of pupils, with errors being focused on the remainder. The meanings ascribed to the statements being judged often appear to be based on the cueing supplied by the physical situation: where a statement matches the 'obvious' purpose for which an experiment might be undertaken the experiment will often be judged to fit that statement, even where details of the experiment show it to be logically flawed or even irrelevant. Thus in 'Stamped', where a description of an experiment involving the time between posting and delivery of mail is given, 43 per cent of pupils stated that it would generate information on the frequency of collection from postboxes: an obvious but erroneous possibility. This is perhaps just a special case of situations where the pupil attributes meanings more loosely than the question writer/marker would accept. However generalization is dangerous, and pupils' criteria are not always less stringent than those of 'experts'. The commonest situation in most questions is for pupils (collectively) *both* to assess testable statements as non-testable and vice versa. The latter is, however, more common overall.

Broadly, where errors can be simply interpreted in these variable handling tasks, they fall into three categories. Firstly, there are failures of shared meaning between writer and pupil, both material (referring to individual phenomena) and in terms of question demand (eg imputing a non-standard sense to the request to state 'what should be changed'). Secondly, there are errors involving (apparent) failure to analyse the presented investigation rigorously, where 'obvious' interpretations are accepted. Finally there are errors of logic, in which the structure of an experiment and the relationship of the independent and dependent variable are apprehended in a non-standard manner. It must, however, be stressed that these interpretations of pupils' 'errors' are essentially logical or descriptive categories rather than dynamic accounts. They are obtained by mapping the 'logical' structure of tasks against that of pupils' responses, without an attempt to elucidate the dynamics of pupils' cognitive functioning directly.

Question Type 3: **identify/evaluate operationalization of control/independent variables**. These questions are in fact intermediate between those involving abstract variable handling and those focusing on the details of experimental operations. Thus they present experimental activity in detailed operational form and require responses in this form. However, in order to move from information to required response the pupil must analyse the details of the experiment into independent and control variables form. Errors differ according to whether questions involve fixed responses or prose responses.

In both types the complexity of the experimental situation appears to have a role in determining the most common responses. In a situation with clear simple variables (especially if these are dichotomized or presented abstractly) pupils appear to have no difficulty in applying the basic schema of independent and control variables. In questions such as 'Milk lid' (a prose response question which involves two pans of milk exposed and not exposed to the air) and 'Acceleration' (a fixed response question involving a sequence of balls, of differing masses and diameters, being dropped) logical confusion is rare. Here the errors are lack of precision and detail (in 'Milk lid') or lack of shared meaning (the identification of 'size' with 'mass' in 'Acceleration').

Greater logical complexity appears to lead pupils to focus on pairs of experimental runs in which the given independent variable has been altered and to ignore confounding variables. This can be seen in a question such as 'Dirty T-shirt' where three variables are altered across five trials. This is, of course, in explicit contradiction to the observed high level of understanding of the basic idea of controlling variables elsewhere in the framework. The change may be connected with the shift from controlling variables in the abstract. By contrast, in this question variables including those requiring to be controlled are actively presented in diverse values.

An alternative approach, and one which indicates a more serious mismatch between pupils' and question writer's logic, occurs where pupils compare situations in all of which the independent variable has its maximum value. This is quite a frequent response which has been observed also in the context of **Performance of investigations** and recalls the problems with independent variables in Question Type 2. Clearly the characteristic of the situation being focused on is not being treated as a 'variable' in the normal scientific sense of that word. To gain insight into the logic being operated here would require a more detailed approach based, for example, on interview protocols. The poorest performance is generally obtained on questions which are both logically and operationally complex.

Question Types 5 and 6: In these questions there is a movement away from variables and their handling to the technical details of operationalization. As a result it becomes more difficult to make general statements about the common elements which can be perceived where pupils fail to score. They can be categorized as involving lack of precision or as being insufficiently explicit. Thus a question requiring a response in terms of counting woodlice in various environments may in fact generate the response from the pupil: 'see where they went'. In a question requiring the measurement and comparison of a pair of temperature changes the pupil might, after an implicit assumption that the starting temperature is constant, suggest that 'whichever is warmer at the end is best'. Such responses might be attributed to lack of linguistic skills or material stimulus. However it is noticeable that parallels can be found in **Performance of investigations**, involving qualitative rather than quantitative operationalization or lack of tactical awareness and preplanning.

Questions are constructed so that only a limited range of effective modes of operationalization are possible, and this can be checked through trialling. In the few cases where novel methods of measurement are suggested by pupils they too are usually insufficiently explicit to justify mark allocation.

Failures in these questions, then, were not the result of pupils working outside the overall logic and intention of the given experiment. This again is consistent with pupils' generally high level of appreciation of the basic structure of an 'experiment'. However it is observed that in questions where the stem implies a fairly complex logical structure pupils' inclination was to evaluate the logical design of the experiment rather than focus on operational details. This again recalls pupils' higher level of performance on this aspect of experimental activity. The particular combination of demands for linguistic precision and internalized representations of experimental activity can generate very low scores among the least able, and commonly only 10 per cent or 20 per cent of this group will score any marks at all.

Question Type 7: Only a small number of questions of this type are used. They are highly focused, requiring above all a very precise, extended response, and the absence of both of these characteristics constitutes the main reason for low scoring.

10.4 Summary

Planning and carrying out empirical investigations can be seen as the core of any view of the science curriculum which emphasizes aims in addition to the transfer of conceptual understanding of specific phenomena. Assessment of such activity runs into two main validity related problems.

The first concerns the significance of science as an investigatory activity within the curriculum and its relationship to that activity as undertaken by professional scientists. In educational terms the rationale of the science curriculum must be clearly distinguished from the latter. The extent to which school based activity can usefully be related to the practice of scientists is increasingly questioned. A pedagogically orientated science curriculum will define the role of investigation independently, with the general education of the whole population representing an important criterion of value. The second problem of validity is internal and concerns the sense in which assessment of scientific (*qua* empirical) investigatory activity can take place in a purely written environment. In terms of the APU framework, does **Planning of investigations** have any right to an existence independent of **Performance of investigations**? The science teams have taken the view that it does, and the resulting survey findings on planning have proved of interest in themselves and provided a useful complement to findings from performing practical tasks.

Children's overall performance in **Planning parts of investigations** is one of the best within the APU framework. In terms of 'ability' (as indicated either by APU performance or examination entry) there is little to remark. The structure of collective performance is quite stable across the various dimensions of the pool of items, showing only the expected trends towards increased polarization as linguistic demands (in response mode) increase or overall question facility decreases. These effects are not, of course, specific to this subcategory. Performance shows no significant relationship to gender in these respects.

The pool of questions has been constructed after a logical analysis of the demands of questions, focusing particularly on abstract variable handling and the operationalization (empirical 'setting up') of variables. This analysis of **Planning parts of investigations** leads to the detection of differences in performance. However these cannot satisfactorily be disentangled from the impact of other aspects of questions, particularly their apparent linguistic demand. Moreover, even within the groups of questions which can be identified on this basis, collective performance varies markedly, indicating the impact of factors which are unidentified and perhaps in principle of limited value through being specific to question or pupil-question interaction. Their removal would, it can be surmised, lead to an unrepresentative question pool. However, having stated all of these qualifications, there is some indication that pupils' performances in planning abstract variable handling are substantially better than those in planning the practical details of experimental activity, and that in some cases systematic errors can be detected.

Turning to the origins of poor performance, we can first note that misinterpretation of meanings by pupils, often in very question-specific ways, can result in large drops in performance, particularly in fixed response questions. More fundamental failures to appreciate the basic structure of dependent, independent and control variables are apparently very sensitive both to the complexity of the given experimental situation and to pupils' theoretical expectations of what will occur. The last evidently leads to a failure to deploy potentially available logical schemata in some pupils. Where more systematic confusions occur these can be hypothesized as due to a failure to perceive phenomena as complexes of systematically manipulable variables. Exploration of this area would require further, methodologically different, research. In general the findings from these questions indicate the need for pupils to experience investigatory activity in diverse contexts and degrees of logical complexity if generalized investigatory competence is to be developed.

Pupils' performance in questions dealing with the practical details of experimental activity presents a somewhat more uniform aspect. Here the demands of precise accounts of operations, or of the reasons for such operations, generate much lower standards of performance. Apart from linguistic demands, the need to visualize sequences of operations before being in a position to make any attempt to answer a question may have a part to play here. This appears to be inevitably very dependent on specific experience.

The absence of experience in the type of question used in this category is evident from pupils' frequent attempts to say what will happen or discuss underlying theoretical issues. It may be the intention or expectation that the science curriculum should have an impact on performance in **Planning parts of investigations**. Moreover, it may also be anticipated that investigatory 'skills' are relatively undemanding compared to the conceptual demands of science and that they can be 'picked up' in the course of science experience which is orientated to transmitting conceptual understanding. The APU findings in this area indicate that, as judged by existing standards of performance, these activities by no means represent trivial tasks for the vast majority of children. Other categories indicate that the demands made by the conceptual framework of science are still greater. Nevertheless, performance on the subordinate tasks undertaken in **Planning of investigations**, whether measured by overall subcategory or more detailed scores, shows considerable scope for improvement and thus, it must be assumed, for specific curricular attention.

11

Performing investigations

11.1 Introduction

The original development of the assessment framework required, almost self-evidently, a category devoted to performance of investigations. Yet the place of such activity in the curriculum has been the subject of controversy, since Todhunter made his famous attack on the value of class experiment and Armstrong celebrated 'the art of making children discover things for themselves', to the present day (Todhunter, 1873; Armstrong, 1898; Woolnough and Allsop, 1985). If investigatory 'skills' can coherently and usefully be described, and developed in the classroom, the major question which presents itself in the wider educational framework is the relation of those skills to the accepted conceptual framework of science.

In particular, should such 'skills' represent a major intended outcome of the science curriculum, or are they mainly useful in conjunction with a strategy for developing conceptual understanding? The APU Science teams took the view that investigatory activity had an independent status and need not automatically involve the pupil in understanding science concepts. Moreover, the difficulty of such concepts for many children might prevent their deploying investigatory skills. Thus tasks in this category have generally relied only on everyday knowledge, though this does not preclude the possibility of an approach to assessment of this category involving questions using the concepts of taught science.

The questions used have been in the general form of individual practical work lasting about 30 minutes. They have been constructed along the lines described in Chapter 10, and their main focus has been on establishing the existence and directionality of the relationships between variables.

In an assessment context the underlying methodological issue concerns the reporting of performance across activities which involve diverse phenomena and which require, for successful performance, many different activities. It was necessary to devise a framework describing generalized aspects of tasks and use it in their analysis. This allowed attempts to identify 'skills' deployed systematically across and within tasks. It also allowed studies to be made of the dependence of performance on task-specific knowledge, cues and structure.

Studies have also been made of the relationship between full practical investigations, and questions in isolation or in non-practical forms. The questions used to assess **Performance of investigations** can be seen as the integration of those in the previous categories. Thus performance on activities which are in a logical sense related to components of **Performance of investigations** tasks might be expected to provide useful information on aspects of the latter. A parallel question which has been addressed concerns the extent to which the larger tasks as units may be seen either as replicable, or as qualitatively altered, by transposition from the practical into the written form.

Evidently the burden of assessment needed to be on what the pupil did and not only on the conclusions drawn or data obtained by him or her. It was therefore necessary to address the 'technical' problem of the standardized collection of information on pupils' behaviour during investigations. This involved both relatively routine problems of administration and comparability and more fundamental questions concerning the aspects of that behaviour to be recorded. The latter needed to be closely related to some view of the meaning of scientific investigation in an educational context. Such problems led members of the team to develop a model of scientific investigation as applied to the solving of simple problems. The model (visualising distinct stages in solving such problems) had both a logical and chronological significance, though the balance between these two was not fully explored and it was used rather as a guide in exploring the field and developing questions. An important element in this approach was that of pupil evaluation of each stage: thus the model was cyclic in character (Driver *et al*, 1984, Chapter 6). From this basis the administrative problems were approached through fairly complex checklists of potential pupil behaviour, drawn up after extensive trialling, and through the monitoring of each pupil's work by a trained assessor on a one-to-one basis. This procedure, though expensive, was felt to be the only one appropriate to the problem and has been retained throughout the assessment.

The manner in which the practical tasks were administered has been discussed in some detail in earlier reports (Driver *et al*, 1984, Appendices 1 and 2). Administrators are often asked to function at two levels: generating both 'behavioural checklists' and summaries of various types. The term 'behavioural' here is not, of course, meant to imply that these are other than interpretative in character. Nevertheless, they represent the lowest level of data from which accounts of pupil performance can be constructed. The basic 'unit of activity' within the behavioural checklists was termed the 'trial'. Thus each pupil is interpreted as beginning a new 'trial' when he or she moves to a new value of an independent variable, uses a new piece of apparatus or clearly repeats a self-contained sequence of activity. A pupil's involvement with a task was, therefore, represented as a discrete number of such trials. Within each trial various significant aspects of activity (materials/apparatus and variable values used, accuracy of measurement and recording of results etc) were identified and recorded. The concept of a trial, of course, aggregates a sequence of evidently heterogeneous activities, made more so by the pupil's interaction with apparatus, into a homogeneous unit. Such a process is, however, essential for converting the continuum of pupil activity into something which may be analysed. It was considered by the team to be an appropriate preliminary rationalization of that activity.

Not all tasks have behavioural checklists, but all require some higher level interpretative activity by administrators. This usually involves the administrator in some judgement of the general strategy used by the pupil, the effectiveness of variable handling, data representation and so on. Explicit and standardized criteria were developed. Pupils were asked to record the data they produced, and their conclusions in relation to the initial questions. However, to supplement this, the script frequently required the supervisor to obtain an oral expression of these conclusions, and usually some further enquiry was made as to the pupil's interpretation of the activity undertaken.

Large scale surveying of this category was first undertaken in 1981, and the results published in the second report (Driver *et al*, 1984, Chapter 6). This survey involved mainly an exploration of the difficult practical and conceptual problems in the field of the assessment of investigatory tasks. In particular it set about the task of defining levels of competence within each task, and of using these levels to allocate pupils to overall performance levels. The surveys of 1982 and 1984 (the category was not included in the 1983 survey) were focused more specifically. In 1982 the issues of presentation, planning and reporting were especially addressed (Gott *et al*, 1985, Chapter 10). In addition, this survey was undertaken with a good deal of collaboration with the age 13 team, which lent itself to comparison across ages and joint reporting. The fourth report gave relatively little attention to the category, other than to notice some evidence that shifting contexts appeared to have marked effects on the balance between performance on parallel practical and written questions (Welford *et al*, 1986, Chapter 8 and Appendix 4). In 1984 the focus shifted to the classes of problems undertaken, and a battery of multivariate, measurement-orientated and other questions was used in the survey.

At age 15 six main **'Performance of investigations'** tasks have been reported:

'Survival'	Driver *et al*, 1982, pp92–108
'Woodlice'	Driver *et al*, 1982, pp108–119
'Paper towel'	Gott *et al*, 1985, pp172–181
'Candle'	Gott *et al*, 1985, pp195–206
'Catalase'	present report, Appendix 6
'Springload'	present report, Appendix 6

In addition, other tasks have been trialled. In the following discussion it will be assumed that readers have access to these detailed reports of individual questions.

Some tasks have been reported in a partially standardized form involving the following elements:

— **reformulation of the problem**: here the emphasis was on variable operationalization;
— **planning the experiment**: this was concerned mainly with establishing initial conditions – it can only be judged from a combination of the 'trials' referred to above;
— **carrying out the experiment**: here the focus was on the correct use of instruments and appropriate organization;
— **variable control**;
— **recording data**;
— **interpretation and evaluation** of data and experimental activity.

This type of analysis can be found in 'Survival', 'Paper towel', and 'Candle'. In other cases a more specific aspect of the question was reported. Within these elements it was often found to be possible to set up explicit criteria (based on behavioural checklists) for performance levels.

The approach and techniques of analysis used in **Performance of investigations** have broken new ground in the field of large scale practical assessment, but they have also indicated the formidable problems in such assessment. They have addressed the purely technical problems of consistent and informative monitoring of practical activities. In addition, they have suggested that a rationale of scientific investigation can be generated and applied across a range of experimental situations. Interpretation of the data so generated has proved more difficult. The extent to which this approach allows comparisons to be made across differing tasks remains unresolved. The allocation of pupils to levels of competence referred to previously showed that questions which have formally a great deal in common generate different profiles of group performance across the aspects identified. In addition the profile may change differently for different individuals. The techniques of generalizability theory and question banking are, however, not sensibly available in this field. These issues provide the main focus for the following discussion.

11.2 Pupils' performance in carrying out investigations

No one aspect of any task can be claimed as its core so far as performance is concerned. The components identified above nevertheless allow us to look for failure points and success points in performance. The key areas to be examined are:

experimental design: a major burden of variable-based tasks is on the design of the investigation. This is taken to mean the setting up of the logical combination of 'data points' or 'runs' required to isolate the interaction of the variables of interest. By this strategy, the presented problem is analysed into independent and dependent variables, and these can be reintegrated in the solution of the problem;

control of variables: potential interfering variables are identified and controlled, and irrelevant variables are discounted;

variable operationalization: many of the tasks involve variables which are either noticeably interpretative in character (eg the performances of woodlice) or derived. The latter are frequently rates of changes (eg rate of burning of a candle). Some variables (eg the effectiveness of a material in keeping someone warm) possess both characteristics. The reformulation of variables into measurable or communicable physical realities has been termed 'operationalization'. There is evidence that the most formidable problems are presented within investigations where effective experimental design is conditional on planning or modes of variable operationalization;

recording of data: in general, the recording by pupils of data obtained and especially their interpretation of such records has been viewed as a subordinate element of pupils' approaches to questions. There are technical reasons for this: pupils are interpreting their own data, and this represents such a diversity of starting points that generating groups of any size with comparable data is very difficult. Post-experimental 'interpretation' appears to be a limited aspect of pupil behaviour. A much greater percentage of pupils generate a reasonable conclusion than record their data in a systematic form. Thus it may be that the majority of pupils operate to confirm or deny hypotheses rather than collecting and 'interpreting' data. Nevertheless data recording has been made a significant aspect of all questions reported to date.

relationships between performances on planning, carrying out and reporting of tasks: the main burden of practical questions is contained in the fields outlined above. However, the major significance of the *practical* context of such questions is the immediate tactile and visual stimulus, together with the feedback provided, with the consequent opportunity for evaluation and development. This varies from task to task. Thus certain completed checklists exhibit considerable complexity as individual pupils develop more effective approaches to the requirements of the task. In this section aspects of performance are surveyed in relation to the various written activities which can accompany or be associated with questions in this category. The comparisons which have been made between planning and carrying out whole investigations are also surveyed.

variations in performance across ability bands

Each of the foregoing headings is now discussed in terms of pupils' performance.

Experimental design

A very large majority of pupils can set up the minimum logical structure for an experiment containing one dependent and one independent variable. However, it can be noted that about five per cent of pupils in 'Paper towel' used only one towel and a similar proportion failed to vary the angle in 'Candle'. This suggests that the experimental concept of a variable (a characteristic which can be abstracted from an entity or situation and systematically altered) is not trivial. On moving to problems with two independent variables, such as 'Woodlice', the percentage of pupils able to handle the interaction falls markedly. Only here is the logic of a multivariate situation brought out. In 'Paper towel' as one characteristic alters the other is anticipated to do so, thus presenting a sequence of arguably unique situations. However, in 'Woodlice' it is necessary to operate with situations varying explicitly in two 'dimensions'.

The maximum percentage of pupils potentially performing at or above the minimum design level in 'Woodlice' is 43 per cent. This consists of 21 per cent offering all four environments together and 22 per cent offering one combination at a time. The latter group have obviated the need to offer the variable values together by operationalizing 'woodlice preference' in terms of individual woodlice. It can be noted here, incidentally, that the separation of abstract variable handling and variable operationalization can be difficult in open-ended situations. Additionally, it can be seen that theoretical understanding of, and approaches to, phenomena can have an important impact on aspects of experimental design.

It is perhaps significant that a similar percentage (45 per cent) of pupils can set up the logical minimum for a two independent variable experiment in 'Catalase' (defined here as carrying out an experiment where each independent variable is tested while the other is held constant, and using at least two values of each), despite the fact that:

- the design is established by a sequence of runs;
- the independent variables are quantitative, and offered in open-ended form.

Substantially higher levels of design performance can be read into 'Springload'. Ninety-two per cent of pupils can set up the minimum experiment required (defined as for 'Catalase'). However, the proportion of pupils whose design is consciously constructed is probably much less. In 'Springload' the use of all nine springs by the pupil automatically leads to an experiment perfectly designed in terms of independent variables (and in other respects). However, when asked to suggest how they had designed the experiment only 50 per cent of those whose responses could be analysed achieved a similar performance level. Some further discussion of the relationship between 'behaviour' and its interpretation will be found in Appendix 6.

These results indicate that the basic strategy of an 'experiment' is understood by the majority of pupils. Its application across diverse contexts is variable, but shows some evidence of being consistently affected by identifiable, systematic aspects of tasks, especially the number of independent variables presented. Performance drops markedly as this complexity is increased. However the impact of contingent aspects of tasks is also large. The transferability of performance for individual pupils across tasks cannot, of course, be assessed.

Tasks of greater complexity than that involved in the possession of two independent variables and one dependent variable have not been used.

Control of variables

As has been suggested, variable control can only partially be divorced from the setting up of dependent variables. Control can occur at a very high level, as in 'Survival' (Driver *et al*, 1984, p106) where the majority of the variables recognised in the checklist as being significant were controlled by 70 per cent or more of the pupils. Within 'Woodlice' the percentages operating controls (28 per cent to 54 per cent according to the variable) are signficantly lower throughout, although this may reflect the manner in which 'woodlice preference' is measured. In 'Paper towel' the percentages are generally greater and more diverse (34 per cent to 70 per cent according to the variable). It is evident, however, that in this aspect of experimental design the question of which variables to control and which to ignore cannot readily be separated from the pupils' understanding of the relevance or otherwise of the variables to the problem under investigation.

The fact that control of variables is attempted, to some extent, by almost all pupils in all experiments suggests that this represents a widespread core understanding of what it is to undertake an investigation. It could be argued that it merely indicates that pupils operate in a 'keep everything the same' mode, and that this is supported by the frequent failure of pupils in post-task discussion to specify controls which have, in fact, been operated. However, the variation within pupils' collective performance across and within tasks suggests that this basic appreciation is only functional when triggered by the theoretical model in use. It may be seen as desirable that such triggering occurs, since blanket application of 'keep everything the same' and an inability to distinguish 'irrelevant' variables, would not usually be considered an intelligent approach to investigation. This situation can only be clarified by a study of variation in performance for individual pupils across tasks.

Variable operationalization

Moving from the framework of variable handling and interaction to operationalization might be expected to lead to more heterogeneous performance across pupils and across tasks. As has been implied above, the way in which a form of words such as 'how fast the spring bobs' is reformulated into a measurement might be very dependent on previous experience. Indeed, as 'Springload' shows, an acquaintance with springs via Hooke's Law can override other considerations. (See Appendix 6.)

Adequate operationalization of variables involves the following elements:

- recognition of the need for quantitative approaches where possible, and avoidance of measurement of redundant variables;
- establishment of derived variable values, such as rates of change;
- skills of manipulation and instrument use, in particular the *accurate* use of *appropriate* instruments;
- repetition of measurements;
- appropriate 'extension' in the sampling of variable values where this is controllable (ie use of a suitable range of runs or readings, etc).

Use of quantitative approaches. This appears to present problems for many pupils. When qualitative interpretations are possible, and sometimes when they sensibly are not, many pupils opted to use them. Thus, in 'Woodlice', some 40 per cent of pupils chose a qualitative approach to 'woodlice preference'. In 'Candle', qualitative attempts were made to judge the rate of burning and the angle by 17 per cent and 20 per cent of pupils respectively. In 'Catalase', where direct qualitative comparison is possible, 59 per cent of pupils used such an approach to measurement of the amount of gas generated. In 'Springload' the equivalent figure (for determining 'rate of bobbing') was 56 per cent. In 'Survival' only 44 per cent of pupils measured the volume of water used in the modelling container. In 'Springload' only 18 per cent of pupils systematically measured the spring dimensions, with another 20 per cent undertaking some unsystematic measurement. This can be contrasted with 'Catalase', where the values of the independent variables (the volumes of liver sludge and hydrogen peroxide) were set up qualitatively in only four per cent of cases.

The tendency to use qualitative approaches can probably be attributed to the pragmatism of pupils and their disinclination, except under force of circumstances, to increase the complexity of the experiment. However, it indicates also a failure to appreciate the role of quantitative approaches in science. The idea that 'quantity' needs to be developed from qualitative perceptions of phenomena perhaps receives little emphasis in the school laboratory: quantities are given rather than actively constructed. For many pupils, the idea that quantitative approaches lead to greater reliability, communicability and repeatability seems to have little impact. However, in both APU and other forms of assessment, the extent to which a task is 'open' in this respect ought to be actively considered. The tendency to assess performance as if a qualitative response was explicitly forbidden, when in fact it is not, should be avoided (Entwistle and Hutchinson, 1985).

For some of the tasks it could be argued that the quantification of potentially qualitative variables is made more difficult by complexities of experimental design or rate of change variables. To eliminate all of these aspects would result in tasks which are too trivial to be undertaken in the context of **Performance of investigations**.

Little can be said about operationalizing qualitative variables in **Performance of investigations** as this has been given little emphasis in the questions used. Qualitative variables with a commonsense meaning are generally preferred. However, it can be noted that in 'Woodlice', only about 62 per cent and 40 per cent of pupils respectively were considered to have achieved an even dampness and adequate degree of darkness in the choice chamber. These might be most appropriately considered as the qualitative equivalents of precision in the use of instruments.

Establishing derived variable values. The difficulty of establishing derivative variables such as rates of change appears not to have been a key factor in generating qualitative approaches. The effective setting up of such variables was achieved by the following percentages of pupils in the stated tasks:

'Survival'	52 per cent	(rate of change of temperature)
'Candle'	70 per cent	(rate of burning of a candle)
'Springload'	32 per cent	(rate of bobbing of a spring)

It is striking that the last of these generates a percentage of effective quantitative approaches similar to that in 'Catalase' (32 per cent), despite the much greater complexity of operationalization in 'Springload'. Moreover, it does not include those who interpreted 'Springload' broadly in terms of Hooke's Law. In 'Catalase', while a quantitative approach can involve merely measurement with a ruler, direct qualitative comparison is possible and frequently used. The utilization of qualitative approaches appears to be very dependent upon such contingent aspects of tasks. These figures are, of course, generated from 'behavioural' data, and it is not possible to claim that in the experiments pupils were consciously determining a true derived variable. Endeavouring to measure a rate of change cannot readily be distinguished from measuring of a simple variable combined with control of time. To approach 'rate of change' as a new 'unitary' variable seems likely to involve the pupil in operating at a more sophisticated level. Nevertheless, whichever strategy is involved, it is clear that above 50 per cent of pupils can effectively reformulate two variables in this way when necessary for the purposes of comparison, under optimal circumstances.

Skills of manipulation and instrument use. Use of apparatus and measuring instruments appears as a separate category as well as forming an element of **Performance of investigations**. In general, the standard of accuracy appears high, except in the measurement of time, where diverse performance levels have been recorded. Time is, of course, often measured in more complex situations. In 'Survival' 63 per cent of those who used a clock were recorded as failing to start it within ± 5 seconds of the required time. Only about 17 per cent failed to use a protractor with sufficient accuracy in 'Candle', whereas some 85 per cent of those who timed failed to do so to an accuracy of ± 2 seconds. In 'Springload', 84 per cent of those who used a clock were considered to have used it with sufficient accuracy, whereas in 'Catalase' the equivalent figure was 67 per cent. The assessment of accuracy in this area requires close supervisor involvement. Such diverse results suggest that assessors' judgement of it varies according to task, and thus that it cannot be measured with reliability in the context of **Performance of investigations**.

Instruments are commonly used appropriately, eg to measure suitable quantities of materials. Attempts to measure, for example, tiny quantities of water with large measuring cylinders were not common. In 'Paper towel', some 14 per cent of trials involved attempts to weigh one sheet of paper on a lever arm balance, though it must be noted that no alternative method of weighing was available. In 'Catalase' only 15 per cent of volume measurements were considered to have been made using inappropriate instruments, and here a wide range of instruments was available.

Repetition of measurements. This aspect of tasks also bears on the question of accuracy, and pupils' awareness of it. The extent to which repetition is practicable varies across tasks. In 'Survival', repetition was undertaken by only seven per cent of pupils, whereas in 'Paper towel' two complete sets of measurements were undertaken by 31 per cent of pupils. In 'Candle', some trials were repeated by 23 per cent of pupils, and all trials by seven per cent, while the equivalent figures in 'Springload' were 50 per cent and four per cent. Here the effect of greater *ease* of repetition is demonstrated by the first figure in each case, and the lack of impact on *systematic* repetition by the second.

Extension of variable values. Pupils showed a tendency to use the minimum possible range of variable values. In 'Candle' 37 per cent of pupils used three or fewer values for the angle. In 'Survival' only 28 per cent of pupils used more than two temperature measurements in establishing the cooling characteristics of the models. However, in 'Springload' all nine springs were used by 37 per cent of pupils, and the percentage using all three possible widths and lengths was 68 per cent. The effect of contingent aspects of tasks on the quality of experimentation can be seen again here. When presented with three towels, three springs, three widths and so on, pupils will often utilize all of them. When presented with an open-ended situation, as in 'Catalase', they will equally often utilize the minimum logically consistent with completing the given task. The situation has parallels throughout the various aspects of operationalization.

In general, performances are fairly good under optimal circumstances, ie where *any* attempt at a task sensibly requires good practice or on those variables which are 'structured' by the apparatus available. A substantial majority of pupils in each case will function quantitatively, use appropriate instruments with reasonable accuracy, and will undertake more than the minimum of measurements logically compatible with the solution of the problem. The removal of these conditions results in a marked reduction in the quality of performance. In particular the possibility of qualitative alternatives leads to a marked increase in such approaches. Systematic repetition of measurements occurs very infrequently, except where a short sequence of tests produces self evidently uncertain data.

Recording of data

Recording of data has been given fairly low priority in reporting this category. Nevertheless, pupils were invited to 'make a clear record of (your) results, so another person can understand what you have found out', and pupils' records are best interpreted as their responses to this request. Pupils' efforts have generally been classified into broad categories such as 'tabulated', 'ordered' 'prose' and 'random'. Clearly the first two are generally more appropriate for recording variable-based data. This allows direct comparison between tasks. Percentages of pupils using these approaches within the tasks lending themselves to data recording are shown in Table 11.1.

Table 11.1 *Recording of data while performing investigations*

(Percentage of pupils recording data in the indicated form)

Question	Tabulated	Ordered	Random/prose
'Survival'	21	54	22
'Paper towel'	13	73	12
'Candle'	27	43	28
'Springload'	19	42	31
'Catalase'	10	42	41

A pattern is observable, though with the influence of questions clearly evident. The largest group of pupils in each case order their data, but do not tabulate it so as to reflect the variables in the questions. A question such as 'Candle', which implies a sequence of pairs of variable values, produces a higher than average percentage of tabular records. 'Springload' and 'Catalase', with a more complex variable structure, generate a smaller number of true tables. This is particularly true for 'Catalase' where the variable relationship is less apparent. Clearly the percentage of pupils who will *always* construct tabular data must be low: perhaps 10 per cent. It seems that pupils' approach is question-dependent in many cases.

Relationships between performances on planning, carrying out and reporting of tasks

The influence of planning investigations

The effects of this and other related activities were presented in the joint age 13 and 15 report (Gott *et al*, 1985). In 'Paper towel' a sample of pupils was asked to produce a written plan of their investigation in advance. This resulted in a few specific improvements in performance as between those who had and those who had not had an opportunity to plan the experiment. Thus 19 per cent and eight per cent respectively produced data in tabulated form. Similarly an increase of about 10 per cent was observed in the proportion of pupils controlling the amount of paper and water used. However, in general it was concluded that planning of experiments produced little significant improvement in performance and that the impact of *using* apparatus was very much greater (Gott *et al*, 1985, pp209–210).

Accounts of tasks subsequent to their performance

A comparison between pupils' written accounts of tasks and their actual performance poses methodological problems. In practical environments certain variables which can be ignored in written versions must be set at some value, and can thus lead to 'improvements' in variable control. In a written account variables may be operationalized in a very abstract way. Nevertheless, pupils' responses can be broadly compared. This showed little variation between experiment and account, except a tendency not to indicate the details which have been noted. In addition such accounts were significantly better than written plans of investigations not actually undertaken (see below). Thus, whereas only 22 per cent of pupils set up a functional method of operationalizing the towels' absorbencies when asked to give a prose account of how they would undertake 'Paper towel', 48 per cent could do this in their written account of the task after undertaking it. This compares with 75 per cent actually achieving this in practice. Similarly, whereas only 23 per cent suggested a quantitative method of measurement in the prose version, 40 per cent did so in their account of the task itself. Again, 35 per cent and 68 per cent of pupils respectively identified the need to control the quantity of paper used, compared to 62 per cent controlling it in practice.

These results have significance for the assessment of practical activity where detailed monitoring of pupils' actions is impractical. They indicate that accounts can be used to obtain in some respects an authentic indication of practical performance though the existence of inconsistencies must be noted. This is particularly relevant in the light of the following section.

Performance on parallel written and practical tasks

As noted in Chapter 10, similar structures are required to assess written and practical approaches to entire investigations. The difficulties in making comparisons are similar to those referred to above. Nevertheless, in two tasks, 'Survival' and 'Paper towel' such comparisons have been made. The results indicate that assessment of pupils' practical activity identifies a wider range of capacities than does assessment by written questions. Tables 11.2 and 11.3 compare performance on two of the aspects of investigatory activity to which reference has been made: control of variables and dependent variable operationalization (Driver *et al*, 1984, Tables 7.2 and 7.6). They refer to three different 'versions' of questions presented respectively in 'prose', 'pictorial' and 'practical' forms, or, more accurately, three questions with strong logical and practical similarities but differences in presentation and response mode. The first two present a written and a partly pictorial introduction to the question respectively.

Table 11.2 *Control of variables in practical and written questions*

(Percentages of pupils controlling variables in type of question indicated)

Question	Variable	Prose	Pictorial	Practical
'Survival'	area of material	15	12	64
	cooling time/change in temperature	30	41	75
	external conditions	31	52	62
'Paper towel'	area/mass of paper	24	33	62
	amount of soaking	55	70	77
	dripping	36	37	38

Table 11.3 *Operationalization of variables in practical and written questions*

(Percentages of pupils operationalizing variables quantitatively in stated versions of tasks)

Question	Prose	Pictorial	Practical
'Survival'	12	39	75
'Paper towel'	44	58	72

An improvement in performance on both aspects is evident. The improvement is not, however, quantitatively consistent across tasks or aspects of tasks. The theoretically more complex setting in 'Survival' (which requires a fairly explicit 'conductivity' model of the phenomenon) exhibits a more marked polarization than 'Paper towel'. The impact on operationalization of the cueing in the pictorial version is also greater in 'Survival'. It seems that practical contexts not only tap different aspects of performance, but could also limit the impact of difference in conceptual demand, perhaps by triggering access to a wider range of theoretical models.

Variations in performance across ability bands

The detailed analysis required in **Performance of investigations** has rendered it impracticable to reanalyse questions to produce data of this type. However the two tasks set in 1984 have been so analysed, using examination entry as a surrogate of ability. The data are presented in detail in Appendix 6. Three ability groups were defined, based respectively on those likely to be entered for 6+ O-levels, for some O-levels and for CSE examinations only. It is perhaps inevitable that the generally observed tendency is for the more able to perform better in a fairly uniform manner across the aspects of tasks studied. Nevertheless there is some variation within this pattern, and this is discussed here. The extent to which these comments apply generally rather than to these tasks only is of course impossible to decide at this stage.

A relative shift in favour of the least able was found in the context of the general strategy of setting up an experimental situation: the area of manipulating some variables and holding others constant so as to obtain information. This was particularly the case in regard to controlling 'contingent' variables, ie variables not needed to be manipulated as part of the question. It diminished as the logical complexity of the task increased. Nevertheless, in 'Catalase' nearly 40 per cent and 60 per cent respectively of the two lower ability groups undertook a test of at least one of the two independent variables. This compares favourably with 73 per cent for the most able pupils.

In 'Catalase' and 'Springload' the approach of the lower ability pupils is, however, less systematic in this area, so that they show a relatively greater tendency to test only one of the two independent variables in both written and verbal accounts. A parallel effect was discernible in the control of 'contingent' variables. The less able show a greater consciousness of the need to control the position and method of release of the load in 'Springload', without applying this systematically throughout their investigation (Appendix 6, Table A6.11).

The least able did relatively less well in the setting up and realization of variables, particularly where this demanded well developed preoperational strategy. To some extent these two aspects of tasks (operationalization and general variable handling) are difficult to separate. Thus in 'Springload' the relative ease of operationalization of the independent variables appears to have been important in enabling the less able to perform relatively well in both practical and oral experimental design.

There are parallels with this situation in the performance of the various ability groups in **Planning parts of investigations**. In general it seems that complex operational details present greater difficulties to less able pupils, possibily through the need to construct mental models of operations in advance of undertaking them. It must be stressed however that this effect is superimposed on a generally uniform downward trend in performance moving through ability groups. The extent to which understanding is present, in the absence of the modelling referred to, may perhaps be called into question. However, in 'Springload' the greatest uniformity of performance across ability groups was observed in oral accounts of the design of the experiment *after* undertaking it. Sixty per cent, 62 per cent and 55 per cent respectively of the three ability groups indicated combinations of springs involving at least one non-confounded test of the independent variables (Appendix 6, Table A6.10). This uniformity may be to some extent a consequence of the tendency of the most able to judge correctly that all springs could be used to test each variable and to state that they used all the springs for this purpose (such responses were not allocated to the category of explicit non-confounded tests). Nevertheless it appears that, for the least able, the concrete activity of manipulating the springs represented a form of variable handling which is reflected in subsequent perception of the task. By contrast, in 'Catalase', where the manipulative complexity of the independent variable prevented many of the less able engaging with the task, the response to post-task questioning showed much greater polarization between the ability groups.

11.3 Discussion

The previous section reviewed the results of assessment of **Performance of investigations** and attempted to look at tasks comparatively. The problems of this are clearly visible. This conclusion returns to consideration of questions of the methodology by which it is possible to develop understanding of pupils' practical performance and its assessment.

It is necessary to stress at this point that types of activity under discussion here are highly constrained by the context of national assessment within which APU functions. Short tasks, quickly appreciated by pupils and undertaken individually represent only one mode of practical activity, though perhaps the one most open to school-based as well as national assessment. These constraints must be borne in mind throughout the following discussion.

Unless individual question reporting is to be adopted it is necessary to identify systematic commonalities across questions. These can be constructed in numerous ways, but the most significant for our purposes are:

- at the level of an analysis of tasks in their 'procedural' aspect;
- at the level of the conceptualization of phenomena required.

In addition, it is anticipated that questions' relationships in these respects will be reflected to some extent in pupil performance: otherwise it is not possible to progress beyond reporting individual question performance. Questions in this category have not been intended to draw on pupils' conceptual knowledge of taught science. This represents a limitation on questions which are suitable as assessment vehicles. Nevertheless, the total elimination of conceptual demand is impossible, not to say meaningless. Thus such demands are taken to be question-specific, and contrasted with the 'procedural' aspects which are viewed as generalized across questions. The following discussion explores the extent to which levels of performance might be usefully defined for the 'generalized' aspects of the APU tasks, and whether 'procedural' demands are realistically separable from 'conceptual' demands.

It is not difficult to develop a broad logical analysis of tasks on a variable-handling basis, particularly in a context divorced from a dependence on the conceptual relationships of taught science. The focus of APU reporting has been on experimental design, operationalization and control of variables, and data recording. These classes of activity appear to correspond to some part of what is normally intended by the phrase 'science processes'. Here the need for a twofold approach is most clearly visible in pupils' approach to control of variables. It seems that almost all pupils function with an experimental strategy where relevant variables are held constant. This, it must be assumed, is transferred across tasks, though speculation is possible as to its continued presence in the context of tasks without a 'scientific' cueing. However, when undertaking an investigation, the question of which variables come into this category is not merely 'procedural'. The judgement a pupil has to make depends on theoretical understanding or familiarity with the phenomena involved (Linn, 1980). From this it is possible to decide whether the variable *needs* to be controlled. Such understanding may, or may not, derive from the concepts of taught science. Sensibly, any evidence of control of variables indicates a pupil operating within the generalized framework. Any more sophisticated assessment inevitably involves an assessment of the pupils' specific *physical understanding*. It cannot be suggested that the total number of variables controlled represents a satisfactory indicator of effectiveness, since 'errors' of commission cannot be ignored. Assessment of this aspect of experimental performance ought to reflect this, with perhaps a fundamental or 'trigger' variable taken to be indicative of minimal operation with the idea of 'control of variables'.

In operationalizing variables a similar, if more complex, picture can be suggested. The requirement of quantitative approaches where appropriate is clear cut. *Effective quantitative* operationalization of the independent, control and especially the dependent variable may be suggested as the key criterion for successful functioning in this aspect of investigation. A minimum standard in such quantification is required, applied to appropriate variables within the investigation. It must be borne in mind however that certain, perhaps all, variable choices imply a theoretical modelling of the given situation. It could, be argued that the interpretation of effectiveness of clothing for humans as 'rate of cooling of water' implies a conductivity model of clothing function. Similarly, considerable numbers of pupils interpret 'woodlice preference' at the level of individual woodlice, rate of bobbing as 'average speed of translation of bob' and a paper towel's ability to hold water as 'physical strength when folded'. Effective assessment of this aspect of performance therefore must always imply assessment of variable interpretation and modelling.

The need to measure two data points in order to determine a third adds formally to the difficulty of this aspect of tasks. However, unless an explicit computation is involved, one of the variables can usually be treated as merely requiring control. The problem here is that potentially diverse levels of pupil understanding can be masked by very similar activities. The most complex situation occurs where two variables alter and require measurement during the course of the experimental run in timing experiments. This could be used as a discriminator for tasks. The range of independent variable values used is also significant, particularly when the number of trials, values etc used are under pupil control. This could be quite a generalizable aspect of performance. The achievement of the minimum standard (usually two values of a variable) is implied within experimental design, and is best incorporated there. To detect discontinuities in phenomena at least three variable values would be required, and this might be suggested as the minimum requirement in this context. Beyond this a range of values is acceptable, ultimately reaching a maximum beyond which data collection for its own sake ensues. This constitutes a poorer level of performance.

The effectiveness of putting techniques into operation may depend on experience with specific pieces of apparatus in regard to accuracy in their use, an understanding of the limits of that accuracy and so on. This may be included within generalized skills of experimentation for the more important instruments. The tension between assessing such skills separately and in the context of whole experiments is quite acute here, and there is evidence that reliable assessment requires the former situation.

It might be expected intuitively that experimental design is the most generalized aspect of **Performance of investigations**. It is also fundamental in its implications for a 'scientific' world view. It was suggested earlier that the idea of a variable was not trivial for pupils. The analysis of phenomena into a complex of independent attributes which can be manipulated to answer specific questions, represents a key element in the idea of an experiment. Increasing complexity in this respect has been shown to have a marked effect on performance, reproduced in different contexts. Moreover, objective criteria of design performance can usually be established. The range of possible experimental situations may represent a dimension along which systematic increases in demand may be made on pupils. It must be noted here that the complexity of competent operationalization of the independent variable can lead to differentiation between tasks, as illustrated in the contrast between 'Catalase' and 'Springload'.

It is clear that other characteristics of tasks cut across those discussed here. In particular they may be embedded in the larger question in a highly structured or an open-ended manner. Examples of the first include experimental design and control of load in 'Springload'. Examples of the second are experimental design in 'Catalase' and operationalization of 'material effectiveness' in 'Survival'. In the first case a more or less mechanical progression (using all springs; using any weight repeatedly) will generate a reasonable performance. In the second, conscious planning and carrying through is required. To this can be added the learning potential provided by the tasks. This is evidently much greater in a task such as 'Candle' or 'Springload', where constant modification of activity and assimilation of data and experience is possible, than one where long-term strategy predominates, as in 'Survival'. Clearly such aspects of tasks render comparison difficult.

Evidence has been presented to show that these aspects of tasks have differential impacts according to the performance groups involved. In particular it has been suggested that the least able are particularly disadvantaged by complexities of operational detail. While, in absolute terms, they perform less well on all aspects of tasks, their general understanding of the overall logic of experimental activity is relatively good. Moreover, the interference of complex operational details of variables appeared to undermine understanding of experimental strategy gained by following the question through practically.

11.4 Summary

In developing assessment techniques for this category APU has adopted a view of scientific investigation in an educational context. The implementation of this view, and to some extent the characteristics of the view itself, have been constrained by the practicalities of national surveying. Nevertheless the approach to investigatory tasks identified above represent an embryonic structure for describing and promoting competent performance in this area. In addition the elements defined offer dimensions along which such tasks can be systematically varied and progressively increased in complexity. In conjunction with APU data a backdrop indication of the demands, interpreted as being reflected in pupil performance of these various aspects, has been presented. Thus it can be seen that areas where pupils' performance is particularly open to improvement include:

— design of experiments in multi-variable situations;
— tendency towards quantification of variables;
— extension of coverage of variables.

The question of whether these key scientific skills, and the others identified previously, are explicitly developed by existing science education practice needs to be addressed. This structure may similarly provide guidance for assessment of performance at the individual level.

However, within the APU approach the characteristics of collective pupil performance have always been found to display diversity and sensitivity to the specific combination of characteristics found in each task. While these elements and the performance data based on them can form a basis for development, the second important theme within APU findings must refer to the limits of systematization, and the need for care in attempting reliable estimates of pupils' investigatory skills. Even when performance on individual tasks can be reliably assessed this does not mean that the elements of performance displayed there are generalized: indeed the evidence from collective data is to the contrary. If such assessment activity were to be undertaken it would require reliability studies at the individual level, and careful attention to the communicability of a highly subdivided understanding of **Planning of investigations**. There is another sense in which the 'limits of systematization' can be interpreted. This refers to the tendency for any analytical scheme to devalue the meaning and richness of practical activity as concretely undertaken. We must be particularly wary of the tendency for any descriptive tool (such as the approach above) to appear prescriptive.

The performance of pupils in APU practical investigatory tasks has been dominated by the enthusiasm and basic competence displayed by pupils. When asked whether pupils showed cooperation and interest one assessor replied:

> 'Without exception. Once embarked upon tasks they became really involved and most were determined to solve the problem – no matter how long it took.'

Other supervisors have expressed their regret at the problems and constraints of undertaking such work in the classroom. Systematization for assessment and other purposes is clearly needed, but it is necessary also to tap and channel existing competence and enthusiasm. Only within these constraints can an approach be constructed with implications for the theoretical underpinning and implementation of practical aspects of the science curriculum.

12

Progression in performance from 13 to 15 years

This chapter explores the links in science performance between 13 and 15 year olds. It describes briefly the similarities and differences between the frameworks of assessment in use at the three ages of the APU science survey programme and outlines the extent of commonality of questions used at 13 and 15. A brief comparison of performance characteristics across the science activity categories is made where there are common question pools. For the remaining activities represented in the framework other performance relationships between the two ages are discussed.

12.1 The assessment framework at the three ages (11, 13 and 15)

The assessment framework which operates at ages 13 and 15 is identical in terms of its subcategory composition (as described in detail in Murphy and Gott, 1984), although the degree of overlap between the question pools varies from one subcategory to another (see below, and Table 12.1). There are differences between the age 11 framework and that used with the older pupils. The two subcategories **Reading information from graphs, tables and charts** and **Representing information as graphs, tables and charts** which operate separately at age 11 are combined as **Using graphs, tables and charts** at ages 13 and 15. **Use of apparatus and measuring instruments,** assessed by means of a common fixed test administered to groups of pupils at 13 and 15, is associated with testing of **Performance of investigations** at age 11 and pupils are tested individually. Some types of question have been included within the question pool for **Making and interpreting observations** at age 11 which have been excluded from the age 13/15 common pool in this category. The reasons for this are given in Chapter 7 of this review. The subcategory **Applying science concepts** is divided into three separate question pools – based respectively on Biology, Chemistry and Physics – for the assessment of both groups of older pupils while no such division is operated at age 11. While **Planning parts of investigations** is very largely similarly defined across the three ages, the actual question pool is different at age 11. **Interpreting presented information** is identically defined at the three ages and administered in the same way. However, the pool of questions is different at each age. As with **Using graphs, tables and charts** there is partial overlapping of questions pools between 11 and 13, and between 13 and 15; there is, however, limited overlap between 11 and 15.

12.2 Question overlap between ages 13 and 15

The amount of directly comparable between-age data which can be derived from APU surveys is limited because the light sampling method of selecting pupils for testing is not designed to track individuals or subsamples of the population from a survey at one age to that at another. Comparison between age groups is therefore based on different pupil populations. Although strenuous efforts have been invested by the monitoring teams in rationalising the question pools at all three ages, there remain differences which preclude direct comparison in a number of cases. Rationalisation has been a continuous feature of the work of the teams seeking to eliminate inconsistencies of domain definition, or differences in survey administration between the ages. The process has sought to make explicit the structure of the question pools in order to ensure that the same definitions of question demands, pupil responses and scoring criteria were being operated. This was necessary since questions had been developed with a particular age group in mind. Table 12.1 (p106) shows the degree of overlap between the pools of questions operational in 1984 at ages 13 and 15. In some categories and subcategories, namely **Use of apparatus and measuring instruments, Making and interpreting observations** and **Planning parts of investigations,** the overlap is total and common tests were used in 1984. In the others, while the process of rationalisation has allowed differences to be made explicit, decisions reached about the age-appropriate nature of questions has not significantly increased the overlap between question pools, which remains partial. Furthermore, random question selection reduces actual survey test overlap still further in these latter subcategories since it is unlikely that all the common questions will be drawn in the same year. An extensive comparison is attempted in the following sections for those activities in which a common test has been administered; where overlap is partial or minimal, comparisons are decidedly limited in scope.

Chapters 4 in both this review and in that at age 13 (Schofield *et al*, 1988) show that the organisation and administration of testing has been very nearly the same at the two ages. Slight differences have occurred in the numbers of questions selected from each subcategory; of questions per test package; of test packages used and of pupils receiving each package, but such differences have reflected administrative convenience and have not affected test score generation.

Table 12.1 *Question pool commonality at ages 13 and 15*

Category/subcategory		Degree of overlap	Size of pool: age 13 (number of questions)	Size of pool: age 15 (number of questions)	Number of questions in common
1	Use of graphs, tables and charts	partial	273	216	94
2*	Use of apparatus and measuring instruments	complete	common fixed test		all
3	Making and interpreting observations	complete	97	97	all
4i	Interpreting presented information	partial	214	179	40
4ii	Applying:				
	biology concepts	minimal	100	124	2
	chemistry concepts	minimal	101	124	4
	physics concepts	minimal	126	129	11
5	Planning parts of investigations	complete	225	225	all
6*	Performing whole investigations	partial	some questions in use at both ages		3

* These two areas of science activity are not represented by randomly sampled pools of questions from which tests are drawn, but by a fixed test and sets of individual questions respectively.

12.3 Pupils' subcategory performance profiles at ages 13 and 15

In each survey during the period 1980–1984 the estimated scores of the 13 and 15 year old populations have been higher in the subcategory **Using graphs, tables and charts** than for every other science activity category. They have been at their lowest for the 'concept application' subcategories. In 1984 differences in the mean scores of 13 year old boys and girls reached significance in favour of girls for the subcategories **Using graphs, tables and charts** and **Planning parts of investigations**, and in favour of boys in **Applying physics concepts**. This latter difference was also significantly in favour of boys at the age of 15 (detailed comment has been made on this gender difference in Chapter 4). Indeed 15 year old boys consistently produced higher mean scores than girls on every subcategory except **Making and interpreting observations**. The difference here in favour of girls has been apparent at both ages in every survey to date, and reached statistical significance at age 15 in 1984.

The patterns of relative performance between the countries have showed that at age 13 the mean scores for pupils in Wales are lower than those of their peers in England and Northern Ireland, except in the practical 'Observation' subcategory (where mean scores from Northern Ireland were significantly lower than those from England). By age 15 the pupils from both Wales and Northern Ireland were consistently outscored across the framework of activity categories by their peers in England. In the case of observation 13 year old pupils in Northern Ireland have produced a significantly lower mean score than their contemporaries in both the other countries. Why these differences occur in which pupils in England pull further ahead of their counterparts in Wales and then Northern Ireland as they grow older is unclear, and would bear further investigation.

Regional differences between mean scores of pupils in maintained schools in England at age 13 have been repeated at age 15. Pupils in the South produced significantly higher mean scores across the framework than those achieved by pupils in the North, with the Midlands pupils either equalling those of the North or falling between the two. Similarly, performance differences between English comprehensive schools classified by their type of catchment area at ages 11 and 13 have been repeated at 15. Pupils in comprehensive schools drawing from rural or prosperous suburban areas performed consistently better at both ages than have those in schools drawing mainly from other catchment areas.

12.4 Comparative performance at ages 13 and 15 across the common science tests

As was shown in Table 12.1 complete question overlap at ages 13 and 15 has been achieved in three subcategories where a detailed comparison of performance in the 1984 survey is now presented. Performance comparisons in the other areas of scientific activity are not attempted because of the limited number of identical questions upon which to comment. The exception to this is in **Performing entire investigations** where there are data for three questions analysed in sufficient detail for sensible comparison to be possible.

Using apparatus and measuring instruments

The 1984 common fixed test administered to both age groups has enabled a more detailed comparison of performance to be made across the range of activities appropriate to this category. The overall mean scores across the whole test at 13 and 15 were 45 per cent and 52 per cent respectively.

Reading scales on pre-set measuring instruments

Table 12.2 reveals a general picture of progression with age in the accuracy of reading instrument scales. A comparative analysis shows that the main sources of error were evident at both ages (summarised for 13 year olds in Chapter 6 of the age 13 review report – Schofield *et al*, 1987, and for 15 year olds in Chapter 6 of this report). Large numbers of pupils read the top of the meniscus, counted two-unit minor scale intervals in ones, mishandled decimals (especially to the second place) or made arithmetic errors. As one might expect, the proportion of pupils falling into any one error class was reduced for the older pupils. Even in the cases of reading the ammeter, where the performance difference was greatest, there were no *different* causes of failure at the two ages; fewer 15 year old pupils made the principal errors of counting in ones (not twos) or of misplacing the second decimal place. Presumably growing familiarity with the use of instruments generally, coupled with increasing maturity explains the improved performance of the older pupils.

Table 12.2 *Performance of 13 and 15 year olds reading various pre-set measuring instruments*

(Percentage of pupils reading each instrument scale to within one minor scale gradation – 1984 fixed test)

Instrument	Set Value	Range of tolerance	% pupils within range age 13	age 15	Difference (15–13)
Ammeter	0.24 A	0.22–0.26	6	35	29
Forcemeter	16 N	14–18	62	80	18
Stopclock	437 s	436–438	33	51	18
Thermometer	43 °C	42–44	72	85	13
Rule	303 mm	302–304	43	54	11
Lever arm balance	117 g	115–119	29	40	11
Manometer	18 cm H_2O	17–19	15	22	7
Voltmeter	1.1 V	1.0–1.2	83	84	1
Measuring cylinder	42 cm^3	41–43	50	51	1
Number of pupils			842	612	

The effect on performance of replacing practical equipment by photographs or line drawings has been reported previously for both ages (Schofield *et al*, 1988; Welford *et al*, 1986). It was noted that at age 15 good quality photographs and line drawings produced comparable proportions of pupils giving accurate readings (except where parallax interfered with accurate photography). For the younger pupils the match was not so exact, making it less appropriate to suggest the use of paper and pencil testing of this basic skill at age 13.

Fifteen year olds used units with slightly greater proficiency than did the younger pupils – they used the correct units more often, and generally showed a greater knowledge of the SI notation.

Using measuring instruments

Results for two further questions from the fixed test are given in Tables 12.3 and 12.4. Table 12.3 shows the proportion of pupils at both ages whose responses were within one minor scale division of the expected answer (*and* accompanied by the use of a suitable unit) for the first question. Pupils were required to measure various physical quantities and to record their reading accompanied by appropriate units.

Table 12.3 *Performance of 13 and 15 year olds using instruments to measure various quantities*

(Percentage of pupils giving responses to within one scale division of expected value *and* using suitable unit – 1984 fixed test)

Instrument	Quantity measured	% pupils age 13	age 15	Difference (15–13)
Stopclock	Duration of light flash (4.0 s)	48	69	21
Forcemeter	Force to lift a given mass (5.5 N)	32	48	16
Rule	Extension of a spring (4.0 cm)	15	26	11
Measuring cylinder	Capacity of a beaker (136 cm^3)	4	10	6
Number of pupils		842	515	

The pattern of performance was similar at each age. Pupils found the timing task to be the most straightforward and the measurement of cubic capacity was completed with least success. The gap between the ages was greatest for the timing task on which pupils at both ages did best and at its smallest for the least well done volume measuring activity.

Table 12.4 shows the results for a second question requiring pupils to make measurements. In this question pupils were required to leave behind a specified amount

Table 12.4 *Performance of 13 and 15 year olds measuring out fixed quantities of materials*

(Percentage of pupils delivering quantities to within one minor scale division of expected value – 1984 fixed test)

Instrument	Quantity specified	% pupils age 13	age 15	Difference (15–13)
Rule	47.3 cm paper tape	26	63	37
Lever arm balance	68 g plasticine	53	68	15
Measuring cylinder	55 cm^3 water	52	60	8
Lever arm balance	82 g sand	29	32	3
Number of pupils		842	504	

of a material which was checked for each pupil. The supervisor recorded the actual amount of material delivered.

As with the previous question, the pattern of performance was similar at the two ages. The exception to this was the comparatively poor performance of the 13 year olds in measuring and cutting a length of paper tape. Many of the younger pupils found this to be a difficult task, and were much less accurate in their attempts to cut a piece of paper tape of exactly 473 mm in length.

The same picture has emerged from each of the other questions used in the fixed test. Fifteen year olds have worked with greater accuracy and have demonstrated greater proficiency in their use of equipment. The performance gap has usually been small, and in some cases very small indeed. In a question based on the use of a hand lens and microscope, for instance, there was almost no difference at the two ages.

As with the reading of instrument scales the main sources of error shown by 13 year olds were still present at age 15, although the proportion of older pupils making a characteristic mistake was always reduced. Such errors, which have been described for both ages, include cutting paper tape obliquely, mishandling arithmetic, failing to subtract readings (initial and final readings), and failing to allow time for a thermometer to equilibrate.

Estimating

In school science pupils are frequently called upon to make estimates. They may need to select appropriate instrumentation for measurement, or they might be required to visualise the amounts of physical quantities in suitable units. Just how well they have been able to estimate under APU test conditions has been assessed using two different types of question. In one, pupils have been asked to estimate the magnitude of various physical quantities by judging their value; in the second, they have been required to produce specified amounts of material. In neither case have they had access to measuring instruments.

Table 12.5 shows for the first type of question the proportion of both groups of pupils (aged 13 and 15) who gave estimates to within 50 per cent of the given value for the physical quantities listed.

Table 12.5 shows that the relative success (proportion of pupils within a certain range of accuracy of value given) was very nearly the same at the two ages, for most of the physical quantities being estimated. Thirteen year olds could estimate length as well as, and mass as badly as, older pupils. Interestingly, although in our experience no comparable quantitative performance data exist, groups of 'expert' adults have fared little better on these estimating tasks!

Table 12.5 *Performance of 13 and 15 year olds making estimates of physical quantities*

(Percentage of pupils making estimates to within 50 per cent of given value – 1984 fixed test)

Quantity	Value	% pupils age 13	% pupils age 15	Difference (15–13)
Volume of water	300 cm^3	11	29	18
Area of shape	200 cm^2	21	30	9
Temperature of water	30 °C	43	50	7
Volume of water	25 cm^3	26	33	7
Mass of block	5,000 g	13	18	5
Time of roll	20 s	75	78	3
Circumference of wire ring	30 cm	73	76	3
Area of bottle base	3 cm^2	50	53	3
Force to stretch band	20 N	34	36	2
Mass of wire ring	20 g	28	30	2
Length of rod	30 cm	85	85	0
Number of pupils		842	508	

Over the fixed test as a whole there was little change with age in the pattern of performance relating to either the gender or the ability of pupils. Both 13 year old and 15 year old girls were outscored by the boys, but the gap between them was no larger at 15 than at 13. The largest performance differences between boys and girls were detected at both ages in questions involving electricity; again the size of the difference was not perceptibly changed with age.

While there was a significant fixed test mean score difference at the two ages (45 per cent and 52 per cent at 13 and 15 respectively) it was not large, and the overlap in performance between the two groups of pupils was extensive. The score distribution at age 15 was simply shifted slightly to the right on the score axis in comparison with the distribution at age 13.

Making and interpreting observations

The overall 1984 mean scores at ages 13 and 15 were 37 per cent and 50 per cent respectively for this subcategory. At both ages the performance of girls was better than that of boys, the difference reaching statistical significance at age 15.

Table 12.6 (p109) shows the mean scores for the different types of question used to test observation at both ages. As was noted in Chapter 7 when age 15 performance in this subcategory was discussed, the individual question mean scores show a wide range of values. The same is true for the scores of the younger pupils.

The pattern of mean scores for the different types of question was similar at the two ages; questions involving classification of objects or photographs and those requiring explanations of scientific events were least well done by 13 and 15 year olds. Questions asking pupils to describe similarities and/or differences, or make predictions consistent with observed data or identify patterns in changes were among the highest scoring at both ages.

Table 12.6 *Performance of 13 and 15 year olds on six different types of question in 'Making and interpreting observations'*
(Unweighted mean percentage scores – 1984 survey)

Question type	Number of questions	% mean score	
		age 13	age 15
Matching objects to drawings	5	33	61
Describing similarities and differences	12	41	53
Making a record of change	3	31	43
Making predictions consistent with observed data or identifying patterns in changes	16	41	51
Classifying	6	32	39
Explaining observations of events	3	31	37
Number of pupils		761	548

The most marked difference between the scores at the two ages was for questions which required pupils to match (and justify the match) given objects to one of a set of similar drawings, although only five questions are available for comparison. In this group of questions the younger pupils were not far behind the 15 year olds in making the match, but did not score well when asked to describe the features or attributes which they had used in matching an object to one of the set of given drawings.

In questions which required identification and/or description of relationships or patterns, pupils at both ages were quite efficient at identifying both the dependent and independent variables operating in an event. In relating the relevant variables the older pupils more often went on to generalize the relationship. More of the 13 year olds stopped short of generalising the pattern, and described the relationship in intermediate, specific terms. They more frequently used phrases like 'the heaviest', 'the slowest' whereas more of the older pupils used general terms like 'the heavier' and 'the slower'. Similar findings have emerged from the two or three common **Interpreting presented information** questions involving variable relationships.

At both ages performance differences between boys and girls at the individual question level did not relate to type of question but were closely allied to question content. The questions which showed the largest mean score difference between boys and girls at age 13 were the same as those at age 15 (as described in Chapter 7 of this report). While the number of questions relating to any one specific area of content is always small the largest margins of difference where boys at both ages outscored girls were on three questions involving electricity. The largest differences in favour of girls, again at both ages, were for four questions about chromatography. The gap between the sexes would seem to have widened with age on these particular questions, whereas no such trend was detectable in other content areas.

A comparison of the most able and the least able groups shows a difference at the two ages. At age 15 the mean score obtained by the top fifth of pupils was 59 per cent and that achieved by the least able pupils was 35 per cent. At age 13 the amount by which the most able outscored the least able was far more variable and the mean scores for the two bands were 54 per cent and 19 per cent respectively. Thus while the scores of the most able increased between the two ages, those of the least able had increased very much more markedly for almost all the questions used in 1984.

Unfortunately, too few overlap questions occur in the category testing **Application of science concepts** to enable sensible comparison to be made with **Making and interpreting observations**. However, where common questions have been used at the two ages the same pattern exists: a narrowing of the gap with age between the bottom and top performers. Presumably the intellectual growth of the lowest performers accounts for at least part of the narrowing of the gap between them and the most able as they progress from 13 to 15. As these less able pupils grow older they become better able to cope with the abstract scientific concepts involved in theory-laden 'observation' questions. It might also just as well be the case that limiting the conceptual load in APU questions used at age 15 to only those concepts encountered by the end of year three has applied a 'ceiling effect' to the most able, so allowing the lowest performers at 13 to show the most progress over the two years up to 15.

Planning parts of investigations

The overall 1984 mean scores at ages 13 and 15 were 32 per cent and 42 per cent respectively. The 'profile' of performance across the types of question used was similar at the two ages; 15 year olds were slightly ahead on mean scores in every case. There were no significant differences between boys and girls at either age.

Performance at the two ages was previously compared in the report of the 1983 survey at age 15 (Welford *et al*, 1986). Fifteen year olds were ahead on all of the seven types of questions used in **Planning parts of investigations**, as they were again in 1984. The pattern at the two ages was the same, with pupils at both ages achieving significantly higher mean scores on question types broadly described as variable handling than they did on those which asked about the operational details of an investigation.

In Table 12.7 (p110) the pupil scores are broken down according to these two main types of question (as described in Chapter 10).

While the older pupils were better at both types of question there was more improvement with age in the variable handling questions. It is possible to speculate about the reason for the higher mean score in variable handling. It may be comparatively unproblematic to

Table 12.7 *Performance of 13 and 15 year olds in planning questions*

(Mean percentage scores on two main groups of planning questions – 1983 and 1984 surveys)

Question type	Number of questions	% mean score	
		age 15	age 13
Operational details	49	30	24
Variable handling	51	54	41

recount a list of factors which might affect the way a given system behaves. It is also probable that the ease with which such activity could be undertaken would increase with exposure to all types of practical work in science. On the other hand it is much less easy to visualise the operational details of an investigation in the absence of an opportunity to try the system out. Performance is perhaps more closely tied to width of experience and recall of specific situations. The relatively slight improvement on 'operational details' questions might thus reflect limited range of exposure to such activity, while performance on 'variable handling' might be less sensitive to lack of width in curriculum experience.

An analysis of the scores of the most able and least able groups at both ages on the two types of questions shows that the least able pupils improved by rather more than the most able. This facet of improvement was more marked for variable handling questions. If variable handling skills are held to be more closely associated with general intellectual growth than operating with the details of specific experimental activities, then the effect observed could suggest that intellectual growth is of greater significance than the taught science curriculum in generating improved performance in this area. The relative contribution of each remains an open question. This is consistent with the argument of the previous paragraphs, and with the view that for these two reasons the science curriculum has comparatively little effect on investigatory skills.

Performing entire investigations

Comparable data are available for three questions which have been used at both ages 13 and 15: 'Woodlice', 'Paper towel' and 'Survival'. While there are some variations in performance characteristics, the overriding impression is of similarity.

In 'Paper towel' the distribution of the techniques of measurement used by pupils was similar at the two ages as was that for the conclusion which they drew (Gott *et al*, 1985, pp179–181). Among the differences that were observed was the fact that 13 year olds showed a greater tendency to control *all* variables, perhaps indicating less discrimination between relevant and irrelevant factors. This may have demonstrated greater enthusiasm at age 13, or perhaps a less sophisticated understanding of the physical phenomena involved. Fifteen year olds made greater use of tabular records of data, and they showed a greater tendency to reconstruct their investigation radically after a period of time (14 per cent of 13 year olds; 34 per cent of 15 year olds). The reconstruction usually effected an improvement. This latter finding may be considered encouraging because it again demonstrates that older pupils showed greater critical awareness and reflection.

In 'Survival', improvement was observed in the operationalisation of 'rate of cooling'. At ages 13 and 15 respectively, 44 per cent and 56 per cent of pupils measured *both* initial and final temperatures (Driver *et al*, 1984, p101). While this improvement is relatively small, the total percentage of pupils described as performing at the highest level for this aspect of the task improved, as a result, from seven per cent at age 13 to 17 per cent at age 15. Much of this advance can probably be attributed to improved conceptualisation of the cooling phenomena involved rather than the manifestation of any generalised investigatory skill.

'Woodlice' showed least difference between the ages. Both groups of pupils performed similarly on both the design of the choice chamber and the method of operationalising 'woodlice choice' (Driver *et al*, 1984, p115; Schofield *et al*, 1984, p64). The slight differences in types of effective performance (one to two per cent in each case) favoured the younger pupils. However, responses to questioning about measurement of woodlice choices partially contradicted this. While 18 per cent of older pupils suggested a quantitative approach based on more than one woodlouse, only five per cent of the 13 year olds did so.

In some tasks in this category performance levels have been produced to characterise pupil approaches to the investigation (see Driver *et al*, 1984, Appendix 2, p195). The criteria for determining these levels have been applied in identical fashion to the investigations carried out by pupils at both ages. Table 12.8 below shows levels of performance for measurement activity in 'Paper towel' and for overall performance levels in 'Survival', at the two ages.

Table 12.8 *Performance levels of pupils aged 13 and 15 in 'Paper towel' and 'Survival'*

(Percentage of pupils at in each age in level indicated)

Level	% pupils at each level in 'Paper towel'		'Survival'	
	age 13	age 15	age 13	age 15
High 1	43	44	10	25
Medium 2*	34	31	34	30
Low 3	23	25	56	43
Number of pupils	824	905	618	202

*Amalgamation of three original levels of 'Paper towel'.

The ways in which particular explicit combinations of investigative behaviour have been extracted from the checklist and subsequently aggregated have been described in previous reports for the two investigations. (For 'Survival' see Driver *et al*, 1984; for 'Paper towel' see Gott *et al*, 1985.) Thus for 'Paper towel' level 1 includes all those pupils who soaked their paper towels completely or for a set time in a suitable container, and then went on to weigh either the wet towel or the water involved, or who measured the volume of water remaining and/or that squeezed out of the paper towels.

In order to qualify for level 1 in 'Survival' pupils must have tested both materials; used at least one layer of material securely held around a metal can of hot water; taken an initial temperature reading for the water and timed a cooling interval of at least two minutes using a stopclock before recording the final temperature(s) of the cans.

The similarity at the two ages is self-evident for 'Paper towel'. For 'Survival' the largest relative change is in level 1 – the best performance level – so signifying a definite improvement with age. Naturally such a shift is accompanied by a smaller number of older pupils locating in the lower performance levels.

The precise shifts of pupils between these levels (and their dependence on curriculum influences) cannot be judged without tracking the same pupils from one age to the next. Nevertheless the closer relevance of the traditional physics curriculum to 'Survival' can be noted. 'Paper towel' is by comparison a much more 'everyday' activity. Thus it can be argued that the influence of the curriculum, assuming some part of the changes can be so attributed, relates rather to conceptualization of specific phenomena than to general investigatory skills.

12.5 Summary

The overlap between the question pools operated at ages 13 and 15 has been described and presented in Table 12.1. It shows total overlap in **Use of apparatus and measuring instruments**, **Making and interpreting observations** and **Planning parts of investigations**. Elsewhere in the framework overlap is partial or minimal.

Comparison of performances has been made at the two ages for those categories and subcategories where common tests have been administered, or, in the case of **Performing entire investigations**, where common detailed analysis of performance has occurred for a limited number of questions.

While science performance for the common tests was usually better for 15 year olds than for 13 year olds the differences were often small. The patterns of achievement across and within the activities tested, and the common errors produced by pupils, were similar at both ages.

However, at the level of individual questions there were larger differences. For instance in **Use of apparatus and measuring instruments**, significantly larger proportions of 15 year olds gave accurate readings of most pre-set measuring instruments and used most instruments with greater accuracy. The older pupils were also closer to the expected value in their estimation of the magnitude of most physical quantities.

In the subcategory **Making and interpreting observations** one group of questions showed the largest differences between the two ages (although based on only five questions). The requirement to match an object to one of a set of drawings showed the older pupils only slightly ahead, but they were very much better at justifying the match than were the younger pupils.

Within the subcategory **Planning parts of investigations** 15 year olds were better at all the question types, but the greatest improvement with age was in the questions broadly described as 'variable handling'.

In **Performing entire investigations** 13 year olds showed a greater tendency to control *all* variables when experimenting on 'Paper towel' and were less proficient in operationalizing the rate of cooling in 'Survival'. It was suggested that differences at the two ages between the proportion of pupils in the high performance levels – little difference for 'Paper towel', larger difference for 'Survival' – reflected the closer relevance of the physics curriculum to 'Survival'. It may well be that the effect of two years extra experience in science activities is to improve understanding of the specific phenomena and conceptual areas studied rather than to develop generalized investigatory abilities. The evidence advanced in this chapter supports such a view, taking into account the current content of school science and the general approach to practical work. The question of whether the activities tested could indeed be improved by appropriate practices and curricula remains unanswered.

However, for at least some of the subcategories of science activity the effectiveness of the taught science curriculum in promoting increased performance cannot be discounted. The next chapter in this report looks at the influence of studying different science subjects (singly or in combination) on the performance of 15 year olds. It suggests that there is a very real impact of science curriculum on performance across the activities tested. Although comparable curriculum data are not available for 13 year olds, it is not unreasonable to suppose that at least some of the general improvement by the age of 15 is related to science subjects studied.

Enhanced performance in any of the science activity categories might not necessarily be due to improvement in effective performance within the intended emphases of the test questions. Improved linguistic skills, a greater familiarity with formal testing and a greater willingness to engage with the tasks could all play a part. Take, for instance, the tendency to think in generalities rather than in specific terms. Fifteen year olds show an increased propensity to do so in the science tests, and indeed the APU Language surveys have shown that the linguistic skills necessary to use generalities are relatively sophisticated and are more likely to have been developed by older pupils (Gorman *et al*, 1986). Perhaps it is this increased linguistic competence which accounts for the improved performance of 15 year olds on such 'generalizing' questions.

Recognition of the power attached by scientists to 'generalization' is likely to have been explained and illustrated more often to pupils of 15 than of 13. It could be that there exists in the minds of pupils a confusion about generalizing in science and not generalizing in the social sciences which might perhaps have been reduced by the time a pupil reaches the age of 15. After all, exposure to and familiarity with more numerous examples of 'when to' and 'when not to' must be a function of time spent involved in the activities and content of taught science and other school subjects. Presumably a similar argument might be advanced in terms of mathematical competence and the increased tendency for older pupils to engage with the quantitative nature of much experimental science.

Turning now to aspects of progression in performance related to the abilities of pupils, the relatively large improvement achieved by the least able 20 to 40 per cent of the population on some of the activities described (observation, variable handling questions in **Planning parts of investigations**) is worthy of note. The least able 13 year olds may still be the least able grouping at 15 (that is not established), but their ability to engage in some of the activities of science has been shown to be much increased. In the tests of observation an advance from scoring fewer than one in ten of the available marks at age 13 to achieving four, five or even six times that score at age 15 is relatively commonplace for this group of pupils. Such an increase shows that there is potential for real improvement among low achievers during the final years of school, at least in the fields of activity described. While not advocating that this should be made the basis of curricular design (indeed, based on data so far available from only one or two subcategories it would be foolhardy to do so), it must be argued that the impact of perceptible improvement on motivation and self esteem is worth very considerable weight in such design. Other such areas, systematically identified, might form the nucleus of the science curriculum for this group. Maybe in developing this point of increasing performance among the lower achievers it is worth speculating that there might be value in aiming for similar attainment levels for lower achievers in a limited range of science content and/or activities, and allowing students longer to achieve success.

13

Subject take-up and science performance

13.1 Introduction

Chapter 2 has illustrated the curriculum variety which exists in science at this age level in schools. Individual pupils might be studying any one or more of the major science subjects or might be studying some form of general or integrated science. It is tempting to ask to what extent the different science course experiences of pupils influence their 'process' performance in science. Paradoxically, the very *variety* of in-school science experience which prompts the question precludes a straightforward answer.

The science surveys have focused on more than one aspect of science performance. At age 15, for instance, seven different domain-sampled subcategories of science performance are assessed each time. Moreover, no single sample pupil attempts *all* the survey questions representing a particular subcategory. Each pupil undertakes an hour long test package, and will typically attempt between 8 and 20 questions from the same subcategory. This 'multiple matrix' sampling scheme, whilst an efficient strategy for producing population performance estimates, severely constrains the kinds of analysis which might normally be employed to investigate links between performance and pupil or school characteristics. In particular, the usual regression technique is not appropriate for this purpose, and we are forced instead to compare the mean performance scores of pupil subgroups in order to investigate possible performance-characteristic associations. It is for this reason that the variety in the science curriculum of pupils at this age poses problems, particularly as there are strong differences in subject take-up rates among boys and girls, and among pupils of different academic ability.

One *could* simply compare the mean subcategory scores of those pupils studying a particular subject with the scores of those pupils not doing so. But this strategy would be highly unlikely to prove fruitful, for a number of reasons. To give a concrete example, suppose we compare the average performance scores of pupils studying Chemistry with those not studying this subject. What information would such a comparison provide? The answer is little of value.

For one thing, many of those pupils not studying 'Chemistry' as a specialist subject will instead be following a combined science course, or some other relevant science or science related subject, and might therefore be learning some Chemistry while doing so. But more importantly, Chapter 2 has shown that the majority of Chemistry (or other 'main science') takers are very able pupils in terms of general academic ability. Indeed, as Table 13.1 shows, the group of Chemistry takers contains a disproportionately high representation of most able pupils in comparison with the survey sample as a whole (and, therefore, in comparison with the whole population of 15 year olds). The discrepancy in representation of most able pupils will be even more severe between Chemistry takers and pupils *not* studying this subject. A performance comparison between these two groups of pupils would, therefore, have rather little to contribute to any meaningful debate about 'Chemistry influence'.

Table 13.1 *The ability compositions of the samples of pupils studying particular science subjects in England in 1984*

(Percentage of pupils of different abilities studying indicated subject alone or in combination with others)

Subject	Number of pupils	O-levels				CSEs	
		8+	6–7	3–5	1–2	4+	0–3
Chemistry	2,148	38	18	18	12	12	2
Physics	2,571	32	16	19	12	17	4
Biology	3,063	28	15	19	14	19	5
General Science	796	2	2	7	10	38	41
No science	254	7	8	7	13	32	33
Whole survey sample	6,675	20	12	16	14	25	13

Table 13.1 shows that any comparison of General Science takers with other pupils would be equally flawed for very similar reasons – this time it is the least able pupils who are severely over-represented in the subject group.

Even if this confounding of ability and subject choice were not so much in evidence, comparisons on the basis of single subjects would not anyway allow any exploration of the possible 'value added' influence of *combinations* of subjects on subcategory performance. It is, for instance, plausible to speculate that a pupil studying Chemistry *and* Biology might be at a double advantage where certain subcategories are concerned. It would be anticipated that two science subjects taken together will promote subcategory performance more than one subject alone. But the interesting question is whether *particular* combinations are more influential than others in this respect.

Unfortunately, sample size considerations dictate that any investigation of curriculum influence can only include the most common subject combinations. These comprise all the possible permutations of the three main sciences (which between them account for almost two-thirds of all 15 year olds in England), and General Science (a further 10 per cent of pupils). We have here, then, eight different subject combinations which taken together describe the science curriculum of nearly three-quarters of the sample 15 year olds in England. The subject combinations are, with percentages of pupils following them in brackets: Biology, Chemistry and Physics (10 per cent); Biology and Chemistry (8 per cent); Biology and Physics (6 per cent); Chemistry and Physics (9 per cent); Biology (19 per cent); Chemistry (3 per cent); Physics (10 per cent); General Science (10 per cent). All other possible subject combinations (such as, for instance, Biology and Physics-with-Chemistry, or Physics, Technology and Rural Science) were being studied by fewer than 5 per cent of the pupils.

Unfortunately, among those groups of pupils studying these different combinations of subjects the problem of ability differences persists. As Table 13.2 shows, fully two-thirds of those 15 year olds studying all three main sciences in England were those intending to take O-level examinations in at least eight subjects (these would include their sciences). Fewer than one in a hundred of these pupils were from the 'least able' group – taking fewer than four CSEs or no examinations at all. In contrast, among those pupils following a single General Science course, two-fifths were 'least able' pupils and fewer than one in a hundred were 'very able' O-level pupils.

Table 13.2 *The ability distributions among those pupils studying particular combinations of science subjects*
(Percentage of pupils of different abilities taking indicated subjects)

Subject	Number of pupils	O-levels				CSEs	
		8+	6–7	3–5	1–2	4+	0–3
Biology, Chemistry and Physics	678	66	18	11	3	2	<1
Chemistry and Physics	583	32	20	21	13	14	1
Chemistry and Biology	518	23	19	25	17	15	2
Biology and Physics	390	23	18	25	14	17	3
Biology only	1,243	13	12	21	18	28	8
Chemistry only	227	14	12	17	21	28	8
Physics only	663	9	10	21	21	30	9
General Science	796	2	2	7	10	38	41
No science	254	7	8	7	13	32	33
Whole sample	6,675	20	12	16	14	25	13

A sensible strategy would be to compare the performance scores of those pupils studying different subject combinations, after taking some account of their different general academic abilities. One way of doing this would be to focus on those pupils within a particular ability band, and to compare the average performances of pupils *within* this band who were following different course combinations.

The ability bands within which the performances of pupils following different courses will be compared in this chapter are termed 'most able' and 'below average'. 'Most able' pupils are defined as before as those intending to take eight or more O-level examinations (the top group in the original six-group hierarchy). 'Below average' pupils are those who were to take four or more CSE examinations, but no O-levels (the 5th group in the original six-group hierarchy).

Unfortunately, since so few able pupils study General Science, likely to be Integrated Science, and even fewer average ability pupils study all three major sciences it is not possible to compare these particular science backgrounds at all. The only way to advance is to confine comparisons *within* particular broad ability bands, comparing performance scores for those course combinations followed by sufficient numbers of pupils to allow sensible interpretation. Even to allow the second strategy it is necessary to agglomerate data from more than one survey. The findings discussed here are based on the cumulated data from the four annual surveys conducted in 1981–84 (the 8+ O-level pupils were identified only from 1981 onwards). The resulting sample sizes vary enormously, from samples as large as 400–500 in the case of most able pupils studying all three sciences to as small as 20 in the case of most able pupils studying Chemistry only. Most sample sizes are between 50 and 100 pupils within a particular curriculum group for a particular subcategory. Before presenting the findings it is important to identify a crucial assumption which will underpin the validity of any inferences about curriculum influence based on subgroup performance differences. This assumption is that those subgroups of pupils following different course combinations were *equal* in terms of average 'process' ability *before they began their current courses.*

For example, suppose a comparison of the mean 'observation' test scores is made between those able O-level pupils studying Biology and Physics and those able O-level pupils studying Chemistry and Physics. Suppose further that the comparison shows a large difference in mean observation performance. Can we say anything about course influence? The answer is yes, but *only* if we can assume that the performance difference has indeed emerged during the period of these subject courses; in other words, we must assume that a similar difference in the same direction did not already hold *before* these pupils began their examination courses in their chosen subjects. Similar arguments apply to inferences based on differences in mean performance between pupils studying different *amounts* of science (ie different numbers of subjects). The less confidence this kind of assumption inspires the lower will be the validity of respective inferences about curriculum influence. Any such inferences will also be sensitive to assumptions about the nature of the delivered and received experiences of pupils in different classrooms following courses with the same title – for instance, 'Physics' is not the same in all schools.

13.2 Influence of amount and nature of science studied

Table 13.3 illustrates some general patterns in the subcategory performance data for the most able pupils, patterns incidentally which were noted in the very first science survey report at this age (Driver *et al*, 1982, Chapter 8). For instance, the highest or equal highest mean scores are achieved in every case by those pupils studying all three sciences. In general, the next highest mean scores are achieved by those pupils taking any two subjects, followed by those taking a single science course. This is a comfortable finding suggesting that the more time pupils spend studying science the better able they are to produce successful performances on the kinds of test considered here.

Table 13.3 *Performance of* most able *pupils broken down by curriculum background**
(Unweighted mean percentage scores for English pupils 1981–84)

Subcategory	BPC	CP	BP	BC	P	C	B	N
Using graphs, tables and charts	87	87	85	83	85	83	78	82
Interpreting presented information	65	66	61	60	59	60	55	59
Making and interpreting observations	56	53	53	55	46	44	48	50
Planning parts of investigations	65	65	62	61	59	57	54	57
Applying biology concepts	55	48	50	50	44	40	47	48
Applying chemistry concepts	60	59	46	51	46	48	41	44
Applying physics concepts	55	53	48	39	48	38	33	38

* The letters P C B N refer, respectively, to Physics, Chemistry, Biology and 'no science'. Thus, BC denotes course combination Biology and Chemistry, etc.

The fluctuations in this underlying general pattern are quite interpretable as responses to the influence of *particular* subjects on performance in *particular* subcategories. For instance, the differential influence of the major sciences on pupils' 'concept application' performances in respective areas is clear. The data in Table 13.3 indicate that pupils who study the appropriate science subject produce relatively better performance scores than usual when applying concepts in that science area. In other words, the evidence is that studying Chemistry enhances the ability to apply Chemistry concepts, studying Physics promotes the ability to apply Physics concepts and studying Biology differentially develops the ability to apply Biology concepts. This is perhaps the least surprising feature in Table 13.3.

More interesting, perhaps, is the evidence that Physics and Chemistry singly or jointly seem to influence performance when **Using graphs, tables and charts** and **Interpreting presented information** more than does Biology. In contrast, Biology singly, and Biology and Chemistry jointly, influence performance on the subcategory **Making and interpreting observations** more than does Physics.

The data in Table 13.4 indicate a very similar pattern of influence for the below average ability pupils. This table differs from that for the able pupils in that in this case there are no figures for the three subject combination (as mentioned earlier too few pupils of below average ability study all three sciences for such figures to be useful), but on the other hand figures *are* now given for the General Science group. Indeed, perhaps the feature of most import in Table 13.4 is that General Science courses in their present form seem to be singularly ineffective in promoting subcategory performance.

Table 13.4 *Performance of* below average ability *pupils broken down by curriculum background**
(Unweighted mean percentage scores for English pupils 1981–84)

Subcategory	CP	BP	BC	P	C	B	G	N
Using graphs, tables and charts	62	67	62	63	66	59	49	48
Interpreting presented information	41	36	36	40	35	33	26	27
Making and interpreting observations	45	35	44	40	38	40	33	34
Planning parts of investigations	43	37	40	38	41	35	28	27
Applying biology concepts	30	28	30	26	26	27	22	21
Applying chemistry concepts	32	26	28	26	31	22	20	21
Applying physics concepts	33	30	27	30	24	21	21	19

* P C B G N represent, respectively, Physics, Chemistry, Biology, general science and 'no science'.

Having established that different science courses, singly or in combination, do seem to be related to science performance to greater or lesser degrees, the next question which arises is '*How* different are their relative influences?'. Again, this a rather difficult question to answer unequivocally, but it might be possible to move some way towards a response.

First, though, it is necessary to adjust the performance score distributions for the various subcategories, to correct for differences in mean scores and in spread. The subcategory distributions do differ considerably in these respects. An indication of the extent of difference is given in Table 13.5 (p116). The Table shows, for instance, that the mark distribution for the 'data representation' subcategory not only has a higher mean score than those of the 'concept application' subcategories (as shown in Chapter 4), but also that it is 'flatter' than these. In other words, there is greater variation in pupils' raw marks for **Using graphs, tables and charts** than there is in the marks of equivalent random samples of pupils when applying science concepts. To give another example, while the mean score for **Interpreting presented information** is closer to those of the 'concept application' subcategories, its spread matches most closely that of the 'data representation' subcategory.

Table 13.5 *Subcategory mean scores achieved by pupil groups of different academic ability*
(Unweighted mean percentage scores for English pupils 1981–84)

	O-levels				CSEs	
Subcategory	8+	6–7	3–5	1–2	4+	0–3
Using graphs, tables and charts	85	79	73	66	57	45
Interpreting presented information	62	54	47	40	33	21
Making and interpreting observations	53	48	45	39	38	32
Planning parts of investigations	61	53	48	40	34	27
Applying biology concepts	50	43	36	31	25	19
Applying chemistry concepts	53	41	35	31	24	17
Applying physics concepts	48	39	34	29	24	18

The usual strategy for modifying distributions to increase comparability is to standardize them. For the purposes of investigating subject influence, all mark distributions have, within each ability group separately, been standardized to have a mean of 100 and a standard deviation of 15 (an arbitrary convention agreed between all APU monitoring teams). New tables can now be produced, corresponding to Tables 13.3 and 13.4, but this time containing standardized mean scores. The new data for the most able pupils are given in Table 13.6.

Table 13.6 *Standardized mean subcategory scores for* most able *pupils studying particular subjects**
(Unweighted mean percentage scores for English pupils 1981–84)

Subcategory	BCP	CP	BP	BC	P	C	B	N
Using graphs, tables and charts	103	104	100	98	100	97	90	97
Interpreting presented information	103	103	99	99	97	98	93	97
Making and interpreting observations	104	100	101	102	93	90	95	97
Planning parts of investigations	104	103	101	100	97	95	92	95
Applying biology concepts	104	98	100	99	94	90	97	98
Applying chemistry concepts	106	105	94	98	94	96	90	93
Applying physics concepts	106	105	100	93	100	92	88	92

* Raw score distributions of most able pupils standardized: mean 100, s.d. 15.

It should be remembered that while there might be ability or motivational imbalances *between* the different curriculum groups, the samples of pupils *within* each group who attempted the different subcategories were matched in all respects. Differences in the standardized mean scores of pupils in particular curriculum groups across the various subcategories are therefore readily interpretable. It is meaningful, for instance, that the standardized mean scores of those pupils studying both Biology and Chemistry are similar across most of the subcategories, falling noticeably in the case of **Applying physics concepts**. It is similarly of pedagogical significance that the mean scores of the pupil samples studying both Chemistry and Physics are highest for the two appropriate 'concept application' subcategories and lowest for the two 'biological' subcategories **(Making and interpreting observations** and **Applying biology concepts)**.

So, Table 13.6 serves to confirm more clearly what has already been revealed earlier, ie that certain course combinations seem to have greater influences than others on particular subcategories. However, the new presentation does not directly provide a quantification of the relative effects. An indication of the *size* of the relative contributions to science 'process skill' development might be found by considering *differences* in the standardized mean scores of pupils following the different course combinations. For instance, if we subtract the mean scores for those pupils studying Chemistry only from the mean score of those studying Physics *and* Chemistry then we have an indication of the possible extra 'benefit' which has accrued from the additional subject – in this case Physics.

Table 13.7 *Relative contributions of specific subjects and combinations to the subcategory performances of* most able *pupils**

Subcategory	BCP	CP	BP	BC	P	C	B
Using graphs, tables and charts	6	10	5	2	6	4	−2
Interpreting presented information	6	8	4	4	4	4	0
Making and interpreting observations	7	6	9	8	4	3	6
Planning parts of investigations	9	10	8	6	6	4	2
Applying biology concepts	6	4	8	6	3	1	5
Applying chemistry concepts	13	14	6	9	6	9	0
Applying physics concepts	14	16	11	4	12	4	0
Average difference	9	10	7	6	6	4	2

* See preceding paragraph for details about the production and meaning of these figures.

The results of this exercise for the most able pupils are shown in Table 13.7. It might be helpful to indicate further how the figures in Table 13.7 were produced. Consider, for instance, the 'CP' column. The figures in this column were produced in the following way. The standardised mean score for the group of pupils studying Biology only was subtracted from that of pupils studying Physics and Chemistry as well as Biology. Similarly, the mean score for those pupils not studying any science at all was subtracted from that of those pupils studying the two subjects Physics and Chemistry. The resulting difference scores were then averaged to estimate the 'influence' of the combination Physics + Chemistry.

In similar fashion, the 'influence' of *Physics* on subcategory performance can be estimated from the figures in the 'P' column. These figures were produced by subtracting the mean scores for those pupils studying Biology and Chemistry, Biology only or Chemistry only from those of pupils studying these subjects *with* Physics. These were added to the difference between the mean scores of those pupils studying Physics only and those taking no science at all, and the result averaged.

Table 13.7 reflects more clearly those patterns of influence noted earlier, but also allows an attempt at quantification. For instance, performance in the 'concept application' subcategories is quite clearly differentially influenced in the anticipated directions by specialist courses in the appropriate subjects. It seems, in fact, that Chemistry and Physics each provide a performance advantage of around a quarter of common standard deviation across the entire framework *except* in the appropriate concept application subcategories where their influence increases to about two-thirds of a standard deviation. The influence of Biology is rather less, reaching its highest level of a third of a standard deviation for **Applying biology concepts**.

The advantage afforded across the framework by the *combination* of Physics and Chemistry is clearer too. This combination is associated with greater differences in mean scores than are the two-subject combinations involving Biology on all subcategories *except* those which are biologically oriented, *viz* **Making and interpreting observations** and **Applying biology concepts**. The performance advantage accorded by the Physics and Chemistry combination is around half a standard deviation rising to two-thirds of a standard deviation for **Using graphs, tables and charts** and for **Interpreting presented information**, and to a full standard deviation for **Applying chemistry concepts** and for **Applying physics concepts**.

The subject combination Biology and Chemistry is particularly interesting, given its relatively *uniform* influence in all three concept application subcategories.

This same general pattern of course influence holds also for pupils of below average ability. The only difference is that slightly greater advantages are offered by *all* course combinations for this group for the three subcategories least dependent on conceptual knowledge and understanding in science: **Using graphs, tables and charts, Interpreting presented information** and **Planning parts of investigations**.

Table 13.8 *Standardized mean subcategory scores for* below average ability *pupils studying particular subjects**

(Unweighted mean percentage scores for English pupils 1981–84)

Subcategory	CP	BP	BC	P	C	B	G	N
Using graphs, tables and charts	106	108	103	106	105	101	94	93
Interpreting presented information	107	103	103	105	102	100	94	94
Making and interpreting observations	106	102	106	103	100	101	95	94
Planning parts of investigations	111	102	108	102	108	100	94	93
Applying biology concepts	107	106	103	101	102	101	95	94
Applying chemistry concepts	112	103	104	102	107	98	95	96
Applying physics concepts	111	106	103	107	98	96	95	94

* Raw score distributions for below average pupils standardized: mean 100, s.d. 15.

Tables 13.8 and 13.9 present the relevant performance statistics for this lower ability group. An immediate feature in the data is the rather poor showing of General Science, and indeed of Biology, in terms of the average performance differences between those pupils studying these subjects and those not doing so. If these differences can indeed be interpreted as subject influence, then it seems general science courses as presently constituted do not effectively promote the development of the kinds of 'process' skills assessed here. Neither do they appear to develop conceptual understanding to any extent given the evidence of the 'concept application' test differences. While Biology seems to be less influential than Chemistry and Physics for most subcategories, this subject *does* differentially contribute to performance in the two biological subcategories, **Making and interpreting observations** and **Applying biology concepts**. The 'size' of its influence in these cases is around a quarter of a standard deviation.

Table 13.9 *Relative contributions of different subject combinations to the subcategory performances of* below average ability *pupils*

Subcategory	CP	BP	BC	P	C	B	G
Using graphs, tables and charts	13	15	10	7	5	3	1
Interpreting presented information	13	9	9	6	4	2	0
Making and interpreting observations	12	8	12	5	5	4	1
Planning parts of investigations	18	9	15	5	11	2	1
Applying biology concepts	13	12	9	6	5	4	1
Applying chemistry concepts	16	7	8	5	9	0	−1
Applying physics concepts	17	12	9	12	5	2	1
Average difference	15	10	10	7	6	2	1

Again the combination Chemistry and Physics is associated with consistently larger mean performance differences than other subject combinations, and the 'influence' of this combination is stronger and more broadly based than it was amongst the most able pupils. For most subcategories the evidence is that the combination Chemistry and Physics contributes three-quarters of a standard deviation to all subcategories except **Applying chemistry concepts** and **Applying physics concepts**, in which the contribution rises to more than a whole standard deviation.

The subject combination Biology and Physics influences the respective 'concept application' subcategories *and* **Using graphs, tables and charts** more than other subcategories. Interestingly, the 'planning' subcategory shows the largest differences for all curriculum groups involving Chemistry. This was not the case among the most able pupils.

13.3 Implications for gender-related performance differences

Given this evidence of course influence on performance, it is reasonable to speculate that the reasons for girls' performance weakness relative to boys at this age in certain subcategories might lie in their different course experiences – a possibility suggested earlier in Chapter 4. Significant differences in the average performance scores of boys and girls have consistently appeared at age 15 for the subcategories **Using graphs, tables and charts, Interpreting presented information, Applying chemistry concepts** and **Applying physics concepts**. Only in the latter case has a similarly persistent difference also been present in the performance data for younger pupils.

Chapter 2 illustrated the strong polarity in the science course choices of boys and girls at 13+, with the girls leaning towards the biological sciences and the boys towards the physical sciences. Might this explain the performance differences in favour of boys at age 15, some of which are newly emerged?

Table 13.10 *The Biology/Physics polarization in the science curriculum of 15 year old boys and girls*
(Percentage of pupils studying particular subjects in 1984)

	England		Wales		N Ireland	
Subjects	Boys	Girls	Boys	Girls	Boys	Girls
Biology only	8	30	7	22	7	25
Biology and Chemistry	4	12	4	7	2	7
Chemistry only	3	4	2	2	1	3
Biology and Physics	7	5	7	4	5	3
Biology, Chemistry and Physics	13	8	12	12	7	10
Chemistry and Physics	14	3	10	3	9	2
Physics only	16	4	16	4	20	3
Biology overall	34	57	32	49	21	46
Chemistry overall	36	28	31	27	20	23
Physics overall	56	21	50	25	43	19
Number of pupils	3,380	3,300	1,293	1,224	1,283	1,483

Before considering this question, it will be helpful to review in more detail the evidence of the Biology/Physics polarization in the science curriculum of boys and girls at this age. Table 13.10 presents the relevant figures.

Table 13.10 shows clearly that greater proportions of boys than girls follow all those subject combinations involving Physics. For Chemistry and Biology the picture is more variable. It is for 'Biology only' that the greatest discrepancy occurs, with fully 30 per cent of all 15 year old girls in England studying only this subject compared with just 8 per cent of boys. Of particular note is the fact that for the two most influential combinations in terms of process development – *viz* 'Biology, Chemistry and Physics' and 'Chemistry and Physics' – the discrepancies in numbers favour boys.

The overall effect of such differences in the popularity among boys and girls of the various subject combinations is seen in the last three rows of Table 13.10. Two to three times as many boys as girls study Physics either alone or in some subject combination, and up to twice as many girls as boys study Biology. For Chemistry the figures are closer, with more boys than girls studying this subject in England and Wales, but (in 1984 only in the five year survey series) more girls than boys doing so in Northern Ireland.

In addition to this difference in the *nature* of the science studied by the majority of boys and girls, there is also a difference in the *amount* of science studied by the two sexes. Boys do on average take more science than girls, as Table 2.4 in Chapter 2 showed, and the difference is greatest in England. In England and Wales more boys than girls study two or three science subjects at this age, although in Northern Ireland the numbers are closer. On the other hand, more girls than boys appear in the 'no science' group in every country.

In view of these differences in science learning experience it is hardly surprising that some subcategory mean score differences have newly emerged in favour of boys between the ages of 13 and 15, and that others which existed to a slight extent in favour of girls at ages 11 and 13 have by the age of 15 disappeared or changed direction.

The evidence in the survey data is that this indeed might explain the emergence of the performance gaps in the subcategories **Using graphs, tables and charts, Interpreting presented information** and **Applying Chemistry concepts**. No performance gaps are present in these cases among those able or average boys and girls taking the same science course combinations. The 'Physics gap', however, remains. Table 13.11 (p119) provides some of this evidence, in the form of standardized subcategory mean scores for those most able boys and girls studying all three main sciences (the only group containing sufficient numbers of pupils of each sex for comparison to be meaningful).

Table 13.11 *Performance of* most able *boys and girls studying all three sciences*

(Unweighted mean standardized percentage scores for English pupils 1981–84)

Subcategory	Boys	Girls	Diff
Using graphs, tables and charts	103	103	0
Interpreting presented information	103	103	0
Making and interpreting observations	102	106	−4
Planning parts of investigations	104	104	0
Applying biology concepts	105	104	1
Applying chemistry concepts	106	105	1
Applying physics concepts	108	102	6

The marginal differences shown in Table 13.11 in the mean scores of boys and girls for the subcategories other than 'observation' and 'Physics concept application' are non-significant, in this and other curriculum groups. Indeed, even the direction of these small differences changes rather arbitrarily from one group to another. The same is the case for comparisons among below average ability pupils taking particular course combinations.

The gap in favour of academically able girls studying all three sciences in 'observation' confirms a general trend in this direction at ages 11 and 13.

The Physics gap, in particular, is extraordinarily strong and stable. It is evident to more or less the same degree for every curriculum group for the most able and the below average ability pupils, as Table 13.12 shows, despite the very small numbers of pupils in some of these groups (for instance, only 10 boys and 28 girls in the 'most able' Chemistry group). Of particular interest, perhaps, is the fact that this early established relative weakness on the part of girls in Physics is evident even among those able pupils still studying Physics at this age.

The fluctuations in the sizes of the performance gaps shown in Table 13.12 should not be assumed to have any pedagogical importance; these probably result from

Table 13.12 *The broad base of performance differences between girls and boys in 'Applying physics concepts'*

(Standardized mean percentage scores for this subcategory)

	Most able pupils			Below average pupils		
	Boys	Girls	Diff	Boys	Girls	Diff
Biol, Chem and Phys	108	102	6			
Chemistry and Physics	106	102	4	109	102	7
Biology and Physics	105	97	8	106	97	9
Biology and Chemistry	99	92	7	105	101	4
Physics only	103	97	6	107	101	6
Chemistry only	97	91	6	107	92	15
Biology only	89	87	2	100	94	6
General Science				99	93	6
No science	98	93	5	100	93	7

the severe differences in sample sizes involved. The important point is that the physics gap between boys and girls approaches a size of about half a standard deviation. This gap is larger in size than the corresponding gap amongst younger pupils (see Johnson and Murphy 1986), *despite* the fact that these 15 year olds were the most able pupils and were studying O-level Physics. It seems that the early established diffference between boys and girls in this area is never overcome. Little wonder then that so many of the 15 year old girls who were studying Physics claimed to find this subject difficult, as Chapter 3 showed.

13.4 Summary

The wide curriculum variety in science at age 15 raises questions about the relative effectiveness of different subjects and subject combinations in promoting performance in the aspects of science assessed in these surveys. Unfortunately, the nature of the survey design combined with the complexity in subject take-up patterns creates sample size problems which constrain any investigation of this issue. Moreover, no information is available about the subcategory performances of pupils studying different subjects *before* they embarked on their present examination courses. If performance differences between different curriculum groups are to be accepted as indicators of 'subject influence', then the assumption must be made that these groups would not have differed in mean performance on the subcategories concerned before they made their optional subject choices at 13+. The validity of any extrapolation from performance differences to inferences about 'subject influence' will depend on the degree to which this previous performance equality can indeed be assumed.

If the assumption *is* tenable, then the survey evidence is that Chemistry and Physics are more influential than Biology for most subcategories. The exceptions are the 'biological' subcategories **Making and interpreting observations** and **Applying biology concepts**. Even in these cases the relative influence of Biology is less than that of Physics and Chemistry on the 'concept application' subcategories in these subjects. The least concept dependent subcategories, *viz* **Using graphs, tables and charts, Interpreting presented information** and **Planning parts of investigations** are all more sensitive to the influence of Chemistry and of Physics, singly or, better, in combination, than of Biology. This is particularly so among the below average pupils. Indeed, among these pupils the indications are that General Science courses in their current form are singularly ineffective in relation to the kinds of abilities assessed here.

The polarization of girls and boys into the biological/physical sciences at 13+ could be expected to result in gender-related performance differences in some subcategories which were not present earlier. The evidence is that the performance gaps in the subcategories **Using**

graphs, tables and charts, Interpreting presented information and **Applying chemistry concepts**, which are newly emerged at age 15 and in favour of boys, *are* attributable to different subject take-up patterns among boys and girls. However, no such explanation is supported in the case of **Applying physics concepts**. Significant performance gaps in favour of boys have consistently appeared at every age in this subcategory, and the evidence is that this gap *persists* even among those able pupils continuing with Physics to age 15.

14

Summary and discussion of findings 1980–84

14.1 Introduction

This final chapter is intended to provide a concise summary, with some discussion, of the main findings of the five annual surveys of pupils' performance in science at age 15. Section 14.2 deals with the performance of the population of 15 year olds in relation to resource provision and subject take-up and in doing so draws together the findings discussed in full in Chapters 2, 3, 4 and 13.

In section 14.3, performance is summarised on each science activity category in the APU assessment framework (Table 1.1). Subcategories assessed in written mode of testing are considered first, followed by the subcategories assessed by group or individual practicals. Further details of performance on subcategories assessed in written mode are provided in Chapters 5, 8, 9 and 10, and further details of performance on subcategories assessed by the practical mode are to be found in Chapters 6, 7 and 11. Section 14.4 considers some general pointers of the findings for science education.

14.2 Performance in relation to resource provision and subject take-up

APU surveys have revealed a predictably complex interaction between such factors as subject provision, subject take-up and pupil performance. In this section some general comments are made on the provision of science curricula during the period 1980–84. This will be followed by some discussion of pupils' attitudes to science and of the resulting pattern of subject take-up. Finally the general characteristics of the performance results will be outlined, in preparation for more specific comments in section 14.3.

As far as laboratory provision is concerned, it was noted in Chapter 2 that there is no evidence that pressure on laboratory space for examination classes was limiting the availability of laboratory experience to pupils lower down the school. Most schools, comprehensive or selective, had enough laboratories to meet their *current* science teaching needs. It could be, of course, that the extent of science teaching in schools had grown into the laboratory time available. Indeed, it is the case that two-thirds of comprehensive schools had little flexibility in terms of laboratory resources and would find it difficult to accommodate *all* their 15 and 16 year olds for 20 per cent of their curriculum time, as advocated by HMI and others.

In terms of subject provision, the three mainstream science disciplines showed little sign up to 1984 of losing their hold on the curriculum. The alternative was provided by courses such as General Science, Rural Science and Human Biology. This general pattern applied in all three countries with only minor differences. In Northern Ireland the provision of specialist sciences was lower than in the other countries (though it was nevertheless still high), whereas in Wales the availability of subjects such as General and Rural Science was relatively high. These differences possibly reflect differences in school policy and organisation rather than being externally imposed or financially determined.

Such provision constitutes a backcloth against which all aspects of pupils' subject take-up and science performance must be judged. Take-up is determined partly by the formal and informal limitations imposed by schools' policies, and partly by pupil inclination and choice. Most schools in England and Wales required average and above average pupils to study some science, but this constraint was less common in Northern Ireland.

Comparatively few pupils in England (four per cent) and Wales (eight per cent) excluded 'science' from their option choices at 13+, though roughly one-quarter of pupils in Northern Ireland came into this category. Survey evidence is that the proportions of 15 year olds studying no science were in all cases falling year by year. Physics and Biology had roughly equal representation, being studied by just less than half of 15 year olds, while Chemistry was studied by just under one-third. Physics and Biology were strongly favoured by boys and girls, respectively, but the take-up of physical science among girls was growing slowly. General Science and the less widely available subjects, such as Rural Science, were studied by relatively few, and generally less able, pupils. It is not clear to what extent these findings reflect positive pupil choices or the option stuctures made available to them since pupil choice, parental preferences and school option policy are so interdependent. There is very strong evidence of ability-related internal school policies by which pupils were 'channelled' into or away from particular subject areas.

Least able pupils rarely studied a traditional specialist subject. What is evident is that there is still a long way to go if the target of the policy statement 'Science 5–16' (DES, 1985) of balanced science for all is to be attained. A third of the most able pupils studied all three sciences, generally up to O-level and almost a third of *least able* pupils studied a single General Science course. *Overall*, about a fifth of all pupils were studying elements of all three main sciences in one or other of these ways.

Pupils' opinions about their science courses are reported in Chapter 3. In opting for science subjects, interest and supposed job value appeared to be of great importance to pupils. The interaction between interest, performance and perceived difficulty must, of course, be borne in mind. For example, girls, who took up the physical sciences in relatively small numbers, claimed to find the subjects more difficult. While such perceptions might not always be justified, the effect on motivation is surely considerable.

Turning to perceptions of employment relevance, a whole complex of mutually interacting effects come into play. The well established gender stereotyping of occupations is to the fore in pupils' assessment of their appropriateness for men and women, and in predictable ways. This stereotyping applies equally to their own job aspirations. In both cases the effect diminishes slightly as ability increases. Other effects, such as increased judgement of male appropriateness as occupations in closely related fields increased in status, were also evident. Judgements of the relevance of the physical sciences (particularly Physics) to occupations was heavily polarized towards those considered male-appropriate. The combined and mutually reinforcing impact of these various judgements is clearly considerable. The gender stereotyping of occupations, school subjects, 'subject occupation' relevance and job aspirations constitutes a cycle of which the origins are as difficult to trace as the best points for potential intervention. What *is* clear from research findings at younger ages is that this stereotyping emerges very early in pupils' lives.

Pupils' views about some typical scientific applications have also been explored. Perhaps disturbingly, girls appeared less well informed, and less inclined to find out, about the social and political implications of physical technologies. They tended to focus on obviously biological and 'humane' issues even more than boys did on 'physical' aspects, with a consequent likely narrowing of real and self-perceived intellectual competence.

Turning to pupil performance, we note first that the APU science assessment framework was not designed specifically to measure the performance of pupils on specific science curricula. A framework designed for this purpose would focus more on understanding of the concepts of the major sciences. In the main, the APU framework looks rather at generalized activities, in which it can be agreed *any* well designed science curriculum might be expected to promote effective performance.

It is not straightforward to draw conclusions about the inherent relative difficulty of one aspect of science compared with another. This is because performance levels depend on the relative difficulties of the questions and mark schemes used in the assessment, and perhaps it is always possible to modify these one way or the other. However, it is the case that the questions and mark schemes used in these surveys were created with a particular age range in mind, to reflect the kinds of science activity actually being undertaken in schools. In view of this, it is interesting to note that a similar profile of performance levels has consistently emerged in every survey at each age. Performance levels in the concept application subcategories have always been lowest, and those in the subcategory **Using graphs, tables and charts** highest.

As far as performance trends over time are concerned, Chapter 4 noted that there is no evidence in the survey data of any such trends over the survey period. The picture is one of general stability, both in terms of the subcategory performance profile discussed above, and in terms of the relative performances of pupil subgroups such as boys and girls, and pupils in England, Wales or Northern Ireland.

Girls have usually produced slightly higher performance levels than boys when **Making and interpreting observations** at this and younger ages. Boys, on the other hand, have consistently produced higher performances than girls when **Using graphs, tables and charts, Interpreting presented information** and **Applying physics concepts** – this latter also features in the data at younger ages. All these differences are not only consistent and persistent but they also usually reach statistical significance.

There are clear performance differences from one catchment area to another, a declining gradation from 'prosperous suburban' to 'inner city'. Somewhat less understandable are the differences in average performance levels shown by pupils in England, Wales and Northern Ireland. From a position of equivalence at age 11, the latter two countries show a fairly wide ranging, though not uniform, slight decline in relation to England. In part this can be related to differences in the proportions of pupils actually studying science at age 15 in the three countries.

Closer to the immediate learning situation of the pupil is the specific science curriculum experienced. Some attempt has been made in Chapter 13 to examine quantitatively the effects of curricula on pupil performance. It is necessary to bear in mind precisely the diversity of pupil experience which is represented by any crude curricular label. The meaning of such labels as 'General Science' for different schools, and even for different teachers within the same school, can be anticipated to vary considerably. The same applies, perhaps less strongly, to the three main sciences. Nevertheless such labels are the only instruments available for the description and analysis of curricular effects. It must also be recalled that the postulated effects described here depend on the *assumption* that pupils display no such differences in performance prior to following the stated science courses at age 15.

Accepting these limitations, the most straightforward and acceptable finding to have emerged from the surveys is that the more science studied the better is the pupils' performance on APU tests. This applies across the entire assessment framework: the improvement in performance does not occur merely in categories based on the application of science concepts. This finding supports both the validity of the assessment questions and the claims to a generalized influence of scientific activity in the curriculum.

Some attempt has also been made to look at the relative impact of *qualitatively* different curricula and differences in their combined influences. Two major findings emerge as a result of this. The first is that, within any ability group, the influence of Physics and of Chemistry is considerably greater than that of Biology. This applies also to combinations of these subjects, so that the combination of Physics and Chemistry appears particularly effective in promoting enhanced performance on APU Science tests. This polarization in effectiveness must have a particular significance for girls, since in their take-up of science subjects they tended to avoid the physical sciences. It could be, of course, that pupils who possess the science abilities measured in the testing are more likely to choose the physical sciences, and because of related affective factors – motivation, liking, etc – are better able to answer APU Science questions. Whatever the causes of the relationship between the components of the science curriculum and performance on the tests, the evidence would suggest that the structure of the Biology curriculum requires new pedagogic approaches more urgently than do Physics and Chemistry.

The second finding from this exercise has been evidence of a general ineffectiveness of courses in General Science for pupils of below average ability, whose performance in such courses was comparable with that of below average ability pupils studying no science. This is the only ability grouping in which sufficient numbers of pupils were studying General Science for such a comparison to be possible. It is conceivable that ability effects are operating *within* the ability grouping which has been used here, and that pupils studying general science courses were in fact a less able subgroup. Whatever the reasons, these findings suggest that the restructuring of science curricula for below average ability pupils should have high priority.

In general when allowance is made for curriculum and ability differences, gender-related performance differences persist only in **Applying physics concepts** and **Making and interpreting observations**. Among the most able pupils who have chosen to study Physics, girls are still outperformed by boys when applying their conceptual knowledge and understanding in this subject area. In addition, girls' differential curricular choices appear to generate other *group* weaknesses in performances in areas such as **Using graphs, tables and charts** and **Interpreting presented information**. Though these differences disappear when allowance is made for curriculum, they remain real in absolute terms, representing in total a considerable and systematic limitation on the skills and knowledge of half the school population.

The most direct method of establishing the kind of effects outlined above would be by undertaking a longitudinal study based on tracing the progression of individual pupils. Such activity is not practicable within the framework of the APU. However, in a few categories it has been possible to comment on the performance of the age 13 and age 15 populations on identical tests. It was noted above that evidence for curriculum effects on performance was strong, but nevertheless conditional on the assumption that initial performance in the activities identified was comparable for curriculum groups. It is appropriate to point to some of the evidence within cross-age results which make curricular effects appear less clear cut. Thus, for example, it has been found in **Making and interpreting observations** that improvements were particularly marked for the least able pupils relatively independent of the curriculum followed. Findings in **Performance of investigations** suggest that improvements are focused on specific investigations with immediate conceptual relevance to the curriculum. This would undermine claims that a general *investigatory* competence is being developed and suggests that the improvements in investigatory tasks are due to more sophisticated conceptualization of the phenomena involved. Such findings exemplify the fact that categories sometimes show different performance effects and the following section considers each of them separately.

14.3 Performance on the science activity categories

In this section, though the categories and subcategories are treated as fundamentally distinct, the various overlaps and parallels in performance are referred to, where appropriate. Categories are broadly subdivided according to the mode of assessment utilized; ie pencil and paper and practical.

Using graphical and symbolic representation

In this category the central finding is that it is the complexity of the task based on a given representational form, rather than the representational form itself, which most obviously affects performance. In elementary data insertion and extraction tasks pupils of all abilities demonstrate some competence when using line graphs, bar charts, pie charts, tables and so on. It can reasonably be assumed that all pupils have been exposed to each of these representational forms at some stage in their school careers (hence the generalized *basic* competence).

Among most able and above average pupils such competence extends across a wider range of tasks, including, for instance, the complete construction of a coordinate graph. Among less able pupils sensitivity to increases in the complexity of the task, of the data represented, or of technical aspects of the representational form (especially scaling), is considerable.

Competence in the use of 'science specific symbolism' (eg electrical circuit symbols) is highly polarized, with less able pupils being unable to display even a basic competence in many cases. This may be due to innate difficulty, perhaps reflecting the implicit requirement for familiarity with concepts of the relevant scientific disciplines. It could even stem from lack of sufficient exposure in the first instance, or, again, lack of reinforcement through use. Whether these points are worth further investigation depends on the significance attributed to these rather specific competencies for all pupils, but particularly for the less able.

Interpreting presented information

This category and the previous one are quite closely related. The key distinction has been based on the need to perceive, or to operate effectively with, generalized relationships in the data. In practice, pupils often appear to use alternative, more limited, strategies in approaching interpretation questions.

Overall performances on generalizing from given data and predicting new data are approximately similar. Among the average and most able pupils, performance on interpretation questions involving simple induction and deduction is similar to that found in questions on the extraction and insertion of data occurring in the previous section, though marks are lost through lack of precision in language or of articulation. Performance generally falls when pupils are required to handle data based on situations involving more than two variables. This applies particularly when data have to be reordered.

As is often the case, among less able pupils performance is erratic even on 'simple' questions. It shows a characteristic sensitivity to diverse aspects of tasks, but in many cases low ability pupils seem to have little appreciation even of the broadest requirements of the tasks. For example, such pupils often treat data as a general stimulus to comment rather than information to be analysed and utilized in ways specified by the 'demand' of the question. It is appropriate to refer here to the evidence which has been presented suggesting that predictive performance is improved when pupils are previously asked to generalize from data, or subsequently asked to justify their prediction. This effect is relatively greatest for the least able. It may be the case that this improvement, if it can be shown to be generalized, is due to the extra guidance given as to the demand of the question and appropriate strategies for its fulfilment. In any event such strategies for increasing engagement with questions seem worth exploring in order to generate positive performance among less able pupils.

The lowest performance is observed among all groups when a request is made to justify predictions. In part this appears to be due to the fact that pupils adopt pragmatic strategies in making their predictions rather than the 'logical' expert-generated strategies anticipated. Here, and elsewhere in the situations where performance tends to fall, a significant mismatch is often apparent between expert perception of phenomena as a set of independent, controllable variables and pupils' less generalized perceptions and strategies. It is a question for research to investigate what meanings pupils derive from instructions requiring the justification of predictions.

Planning investigations

In the subcategory **Planning parts of investigations** an attempt has been made to establish a pool of questions focusing on the various generalized components of experimental activity. As for the other subcategories, the finding is that performance varies as much within the sets of questions as across them. There is some evidence, though not conclusive, of a polarization in performance favouring questions requiring *variable handling* as opposed to those involving the pupil in the *operational details* of experimental activity. In addition this effect appears to favour the less able who find operational details hard to describe. Pupils' difficulties in expressing the details of experimental activity could of course be merely linguistic in origin, but data on **Performance of investigations** show some parallel effects. It can also be noted here that there is some evidence, from responses in **Performance of investigations**, that complex experimental detail can prevent less able pupils from perceiving the broad purposes and logic of investigatory activity.

Pupils' difficulties in the area of planning operational details are difficult to categorise beyond general failures in precision and articulation of detail. In many cases this appears to be specific to the phenomenon involved and may reflect both a general lack of experience in planning experiments and a more specific lack of experience in the subject matter of the question.

Pupils' difficulties in handling variables can often be more informatively described. At the most general level such difficulties often seem to involve alternative formulations of the rationale of manipulating independent variables. In some cases this stems from a confirmatory approach, in which pupils appear to assume effects consistent with some 'theoretical' expectation and look for situations which will maximize or equalize what is 'expected to be the right answer'! This approach may have its origins in a kind of tunnel vision induced by the highly directed form of 'investigatory' activity commonly undertaken within traditional school science courses. The extreme manifestation of this situation is found among those pupils who respond to questions about the planning of experiments by giving theoretical explanations or accounts of expected results.

Applying science concepts

Explanations and predictions on the basis of theoretical understanding are expected to be undertaken in this category, which is subdivided into the disciplines of Biology, Chemistry and Physics. These subcategories consistently return the lowest scores in the assessment framework.

On comparing performance across the major disciplines the clearest finding is that the average scores in **Applying biology concepts** for pupils following Biology courses are lower than those of pupils studying Physics and Chemistry in the subcategories based on applying the concepts of these disciplines. In consequence the apparent improvement generated from following a Biology course is less than that from the other two disciplines. The effect remains even when allowance is made for general ability. This finding has been related to those of studies on classroom interactions. It has been tentatively suggested that it might stem from differences in these interactions across disciplines, and particularly the more routine, 'fact bound' activity in Biology. This point is particularly significant when the dominant role of Biology in the science curriculum of girls is recalled.

The strongest gender-related difference in performance, in favour of boys, can be found for **Applying physics concepts**. It has been shown that allowance for curriculum and ability has little effect on the differences. It seems likely that diverse factors are at play here, and evidence from other sources suggests that these are operative from an early age.

More detailed studies of performance in particular conceptual areas have been notable mainly for indicating the diversity of response to questions. Various aspects of questions and of the particular phenomena under discussion seem to have marked performance effects. Judgements of pupils' collective understanding of a particular area can shift markedly on moving from one question to another. This appears to have the implications, both for classroom practice and assessment, that diversity of treatment and of phenomena is essential for a generalized understanding of concepts. A conceptually less ambitious science curriculum, which gives pupils time to appreciate the meaning of concepts and opportunities to discuss the application of conceptual knowledge, would also allow greater emphasis to be given to procedural knowledge and skills.

Using apparatus and measuring instruments

Turning now to the 'practical' categories, **Use of apparatus and measuring instruments** represents potentially a focused and well defined body of 'entry skills' to practical science. The basic pattern observed elsewhere, of general competence at a fairly routine level, is repeated. Again, the introduction of complexities has wide ranging effects, though this effect cannot be related so readily to overall ability as in **Use of graphical and symbolic representation**. One parallel with that category, however, is the significance of more complex scaling on measuring instruments and the introduction of decimals. Sensitivity to the physical phenomena involved, and to the overall requirement of the tasks in more complex cases, is also considerable. Finally, pupils displayed a variety of interpretations to the criteria of accuracy appropriate in particular cases.

The implementation of the National Criteria for Science (JCNC, 1985), which require that 20 per cent of marks in GCSE Science examinations be awarded on the basis of practical work, gives added impetus to the acquisition of experimental skills. In the particular circumstances of this category, it is necessary to avoid encouraging arid and featureless measurement activities to the exclusion of testing such 'atomistic' tasks within the context of experimental activities. The attractions of reliable tests should be weighed against those reflecting the validity of such activity. The use of more holistic contexts, despite the problems of reliability this imposes, should be given careful consideration. Indeed, the need for coherent teaching and assessment strategies, which develop measuring skills, will require careful thought. The results from APU surveys of performance on this category form an important starting point.

Making and interpreting observations

Within this category a number of fairly clear findings can be noted. In classificatory tasks pupils were more effective at carrying out the classifications than in expressing criteria for the classifications produced. This effect was particularly noticeable among less able pupils, who were much less successful than the most able pupils in the latter activity.

In describing similarities and differences there were two major limitations on performance. The first was simply lack of detail, broadly a failure to record a sufficient number of observations. The second involved failing to record attributes *comparatively* across instances, but rather stating them individualistically. This aspect of performance illustrates a strand running through the category system, including also **Interpreting presented information** and **Planning and performing investigations**, in which pupils do not systematically perceive or handle the characteristics of phenomena as variables running across instances in the manner of 'experts'. Under certain circumstances many appear to operate with alternative formulations, as indicated for **Planning parts of investigations**, or to treat instances as essentially individual.

The observations, or selections of observations, made by pupils in this category show considerable dependence on the science subjects they studied. This tends to support the view that 'neutral' observation has limited potential as an assessment instrument. It appears that educationally significant tests in this area must recognise and confront the 'theory-laden' character of scientific observation. Thus **Making and interpreting observations** may represent an alternative means of testing conceptual understanding.

In questions where the phenomena in use were systematically altered (with consequent alteration in available observations) pupils performed well in identifying the variables in play. Performance was less good in converting the available data into generalized, potentially causative relationships. Again, as in **Interpreting presented information**, the step to such relationships does not appear straightforward for many pupils.

It can be observed, however, that relatively few pupils who attempted this step limited themselves to referring merely to extreme cases, but expressed the full relationship.

Performing entire investigations

The category **Performing entire investigations** may be thought of as the integration of the practical and pencil and paper activities so far discussed. Performance is of course by no means merely an amalgamation of those referred to previously, and it is difficult to treat each question other than holistically. Certain cross-question and cross-category trends are, however, discernible.

Most positively, pupils' willingness to engage with the practical investigations which have been administered was found to be high. This, rather than any narrowly cognitive influence, seems to have had a central role in generating the much higher performance levels observed in performing, as opposed to planning, investigations. Pupils have demonstrated a basic appreciation of what it means to undertake investigation and to control and manipulate variables.

Their performance on many aspects of the operational details of tasks is less effective, which parallels the situation in **Planning investigations**. Poor performance is common in the use of quantitative approaches, the adoption of suitable ranges of values for the variables being investigated and the systematic repetition of measurements. Pupils have most difficulty where variable operationalization requires a well worked out strategy rather than single measurements of each variable. This is particularly the case in tasks which involve more than one independent variable.

The last point is important in highlighting an important determinant of performance. As the complexity of the independent variables within tasks is increased, performance falls away quite quickly. This may be attributed merely to underpinning levels of cognitive development, but there is some evidence that less able pupils have developed quite a sophisticated understanding of the way in which the variables are working in the task when responding to questions posed orally after carrying out an investigation to its conclusion. When pupils are prevented, by complex operational details, from completing an investigation, less able pupils have performed much worse in relation to the more able.

There is evidence that pupils' subsequent written *accounts* of investigations give an accurate, if selective, indication of their actual performance. Purely pencil and paper accounts without previous practical activity provide little guidance to this. Above all, the diversity of performance in the various aspects of 'investigation', identifiable across different questions indicates the need for similar diversity in any attempt to assess individual pupils reliably in this activity. A related finding is that opportunities to plan investigations, in the absence of interaction with apparatus, are ineffective in generating improvements in practical performance.

14.4 Some general pointers for science education

It is not the function of this review report to speculate in detail on the implications of the findings of the APU surveys at age 15. Indeed, the Department of Education and Science has sponsored an independent appraisal of the findings in science, as in the other curriculum areas monitored by APU, to assist those involved in assessment and examinations, in curriculum development and in teacher training. It does seem appropriate, however, to make just a few general points.

If it is accepted that pupils should be presented with a broad view of science, then it seems important to give emphasis to both conceptual and procedural knowledge, not only in the curriculum but also in the range and scope of assessment methods and techniques. In fact a recurrent theme in this report is the call for diversity and a widening of the range of learning contexts within which pupils may experience the procedures of science and the applications of various concepts within a reduced span of content. This involves increasing the opportunities to investigate diverse phenomena and the opportunities to use skills and procedures in a wide range of situations, together with the assessment of such abilities across the widest practicable range of tasks.

The available evidence on conceptual and procedural understanding suggests that the more able pupils can apply abstract science concepts with some success and that they also have generalized investigatory skills which they can apply, even in some novel situations. At the time of the surveys in 1980–84, many of these more able pupils, roughly a third of all pupils likely to be entered for eight or more O-levels, were following courses in all of the three separate subjects of Biology, Chemistry and Physics. About one per cent of these more able pupils were following courses in General/Integrated Science.

The least able pupils have limited success in answering either concept based or procedural based questions. However, the findings have identified complicating factors in assessment questions which, if avoided, raise the success rate of the lower achievers. In particular, the findings on **Performance of investigations** suggest that if the apparatus is straightforward and the tasks which are set require only limited conceptual understanding or specific language or mathematical skills, then pupils perform well and with enthusiasm.

It would be a mistake to interpret the findings of the APU survey as possibly suggesting that the adoption of a kind of all purpose 'process science' will solve all our problems and particularly that it will provide a general and unproblematic route to 'science for all'. APU evidence supports the view that there is greater scope for producing a rewarding science curriculum based on investigatory science than one which is based on the transmission of concepts from the major science disciplines. But in leading to this view APU activity has clearly shown that the concept of 'science process' is itself highly sophisticated in respect of the activities to which it refers, the complexity of pupil behaviour when undertaking 'procedural' activity, and the problems displayed by less able pupils when undertaking more complex activity of this kind. The devising of tasks which provide pupils with opportunities to learn both procedures and concepts is a formidable challenge for curriculum developers.

All the category chapters in this report have pointed up the problems of defining 'process' for assessment purposes. The same problems must exist for determining the aims and content of a 'process'-led curriculum. This applies particularly to curricula which must also undertake a treatment of phenomena using concepts from the main disciplines. In terms of assessment, and the picture of performance which results, APU has demonstrated that details of the construction of the questions in different categories, the demands made and the marking schemes used all make major contributions to the overall complex picture. Important techniques have been developed to assess population performance on the science activity categories identified in the APU framework. How such techniques might be applicable to the reliable and valid assessment of individual pupils is a question for further research and development.

The attention of teachers and science educators generally has recently been sharply focused on the assessment of practical work. Much of the APU science framework lends itself to consideration in this sphere. Innovative assessment questions have been designed to describe pupils' investigatory and observational behaviour, while findings from the experience of testing the basic experimental skills have pinpointed both common errors and problems associated with testing a diversity of activities. Questions of this kind, and pupil performance on them, are of particular relevance in the context of in-school assessment of practically related abilities and the curricular feedback which it must involve. The consequences of the diversity of response observed must, however, be noted. It implies that *reliable* assessment of individuals will prove very demanding of teacher, pupil, and laboratory time. Findings of the surveys using such questions have stressed many positive achievements of pupils, not least being the enthusiasm with which pupils at the three ages tackle practical tasks, the relatively large numbers who achieve a basic competence, and the confidence which success engenders.

When the APU was set up in the Department of Education and Science in 1975, science was selected as one of the key areas in which a comprehensive programme to monitor the performance of pupils was to be established. In its Consultative Paper 'Assessment of scientific development' (DES, 1977) the Science Working Group concluded by saying that one of the important aims of the programme was that 'published results shall make a useful contribution to the development of science education'. If those concerned with policy issues and assessment, including science advisers, heads and heads of science departments in schools and colleges, senior examiners, teacher training tutors and researchers in assessment and science education, find that this review of APU survey findings 1980–84 helps to inform discussion, to identify issues and to provide 'food for thought', then one of the stated aims of the Science Working Group will have been achieved.

References

ARMSTRONG H. E. (1898). 'The heuristic method of teaching, or the art of making children discover things for themselves'. Board of Education, *Special Reports on Educational Subjects, ii,* 389–433.

BELL B. and BROOK A. (1984). *Aspects of secondary students' understanding of plant nutrition: full report.* Children's Learning in Science Project: Centre for Studies in Science and Mathematics Education, University of Leeds.

BROOK A. BRIGGS H. and BELL B. (1983). *Secondary students' ideas about particles: summary report.* Children's Learning in Science Project: Centre for Studies in Science and Mathematics Education, University of Leeds.

BROOK A. and DRIVER R. (1984). *Aspects of secondary students' understanding of energy: full report.* Children's Learning in Science Project: Centre for Studies in Science and Mathematics in Education, University of Leeds.

CAMBRIDGE INSTITUTE OF EDUCATION (1985). *New perspectives on the mathematics curriculum.* London: DES.

COMBER L. C. and KEEVES J. P. (1973). *Science education in nineteen countries.* Stockholm: Almqvist and Wiksell.

DEPARTMENT OF EDUCATION and SCIENCE (1977). *Assessment of scientific development.* London: Assessment of Performance Unit.

DEPARTMENT OF EDUCATION and SCIENCE (1979a). *Science progress report 1977–78.* London: Assessment of Performance Unit.

DEPARTMENT OF EDUCATION and SCIENCE (1979b). *Aspects of secondary education in England.* London: HMSO.

DEPARTMENT OF EDUCATION and SCIENCE (1980). *Girls and science.* HMI Matters for Discussion 13. London: HMSO.

DEPARTMENT OF EDUCATION and SCIENCE (1985). *Science 5–16: a statement of policy.* London: HMSO.

DRIVER R., CHILD D., GOTT R., HEAD J., JOHNSON S., WORSLEY C. and WYLIE F. (1984). *Science in schools. Age 15: Report No 2.* Report on the 1981 APU survey in England, Wales and Northern Ireland. London: Assessment of Performance Unit.

DRIVER R., GOTT R., JOHNSON S., WORSLEY C. and WYLIE F. (1982). *Science in schools. Age 15: Report No 1.* Report on the 1980 APU survey in England, Wales and Northern Ireland. London: HMSO.

DRIVER R., HEAD J. and JOHNSON S. (1984b). 'The differential uptake of science in schools in England, Wales and Northern Ireland'. *European Journal of Science Education, 6,* 19–29.

EGGLESTON J. (1983). 'Teacher-pupil interactions in science lessons: explorations and theory'. *British Educational Research Journal, 9,* 113–127.

ENTWISTLE N. and HUTCHINSON C. (1985). 'Question difficulty and the concept of attainment'. In Entwistle N. (Ed) *New Directions in Educational Psychology.*

ERICKSON G. L. and ERICKSON L. J. (1984). 'Females and science achievement: evidence, explanations and implications'. *Science Education, 68,* 63–89.

FAIRBROTHER R. W. (1978), 'Assessment of practical work'. In Jones J. G. and Lewis J. L. (Eds) *The Role of the Laboratory in Physics Education,* ICPE, GIREP. Birmingham: John Goodman and Sons (Printers) Ltd.

FOXMAN D. D., BADGER M. E., MARTINI R. M. and MITCHELL P. (1981a). *Mathematical development, Secondary Survey Report No 2.* London: HMSO.

FOXMAN D. D., CRESSWELL M. J. and BADGER M. E. (1981b). *Mathematical development. Primary Survey Report No 2.* London: HMSO.

FOXMAN D. D., MARTINI R. M. and MITCHELL P. (1982). *Mathematical development. Secondary Survey Report No 3.* London: HMSO.

FOXMAN D. D., RUDDOCK G., JOFFE L., MASON K., MITCHELL P. and SEXTON B. (1985). *A review of monitoring in mathematics 1978–82.* London: Assessment of Performance Unit.

GAMBLE R., DAVEY A., GOTT R. and WELFORD G. (1985). *Science at age 15.* Science Report for Teachers: 5. London: DES.

GORMAN T. P., WHITE J., ORCHARD L. and TATE A. (1983). *Language performance in schools. Secondary Survey Report No 2.* London: HMSO.

GORMAN T. P., WHITE J., BROOKES G. and KISPAL A. (1987). *Review of language monitoring 1979–1983.* London: HMSO.

GOTT R. (1984). *Electricity at age 15.* Science Report for Teachers: 7. London: DES.

GOTT R., DAVEY A., GAMBLE R., HEAD J., KHALIGH N., MURPHY P., ORGEE T., SCHOFIELD B. and WELFORD G. (1985). *Science in schools. Ages 13 and 15: Report No 3.* Report on the 1982 APU surveys. London: Assessment of Performance Unit.

Hannan D., Breen R., Murray B., Hardiman N., Watson D. and O'Higgins K. (1983). *Schooling and sex roles: sex differences in subject provision and student choice in Irish post-primary schools.* Dublin: The Economic and Social Research Institute.

Harlen W., Black P., and Johnson S. (1981). *Science in schools. Age 11: Report No 1.* London: HMSO.

Harlen W., Black P., Johnson S. and Palacio D. (1983). *Science in schools. Age 11: Report No 2.* London: Assessment of Performance Unit.

Harvey T. J. and Edwards P. (1980). 'Children's expectations and realisations of science'. *British Journal of Educational Psychology, 50,* 74–76.

Helwig J. T. and Council K. A. (1979). *SAS users' guide.* Raleigh. N. C: SAS Institute.

Holding B. (1985). *Aspects of secondary students' understanding of elementary ideas in chemistry: summary report.* Children's Learning in Science Project: Centre for Studies in Science and Mathematics Education, University of Leeds.

Hueftle S. J., Rakow S. J. and Welch W. W. (1983). *Images of science.* University of Minnesota: Science Assessment and Research Project.

Johnson S. and Bell J. F. (1985). 'Evaluating and predicting survey efficiency using generalizability theory'. *Journal of Educational Measurement, 22,* 107–119.

Johnson S. (1988). *National Assessment: the APU science approach.* London: HMSO.

Johnson S. and Murphy P. (1986). *Girls and physics. Reflections on APU survey findings.* APU Occasional Paper No 4. London: Assessment of Performance Unit.

JCNC (1985). *GCSE: the National Criteria for science.* London: HMSO.

Kahle J. B. (Ed) (1985). *Women in science.* The Falmer Press.

Kelly A. (Ed) (1981). *The missing half. Girls and science education.* Manchester University Press.

Kelly A. (1981). 'Choosing or channelling?'. In Kelly A. (Ed) (1981). (Above.)

Kerslake D. (1981). 'Graphs'. In Hart K. (Ed) *Children's Understanding of Mathematics: 11–16.* London: John Murray.

Lie S. and Bryhni E. (1983). 'Girls and physics: attitudes, experiences and underachievement'. In *Contributions to the second GASAT Conference.* Oslo: University of Oslo, Institute of Physics.

Linn M. C. (1980). 'When do adolescents reason?' *European Journal of Science Education, 2,* 429–40.

Murphy R. J. L. (1982). 'Sex differences in objective test performance'. *British Journal of Educational Psychology, 52,* 213–219.

Murphy P. and Gott R. (1984). *Science assessment framework age 13 and 15.* Science Report for Teachers: 2. London: DES.

NAEP (1978a). *Science achievement in the schools. A summary of results from the 1976–77 National Assessment of Science.* Washington: Education Commission of the States.

NAEP (1978b). *Three national assessments of science: changes in achievement, 1969–77.* Washington: Education Commission of the States.

NAEP (1979a). *Attitudes towards science.* Washington: Education Commission of the States.

NAEP (1979b). *Mathematical knowledge and skills.* Selected results from the Second Assessment of Mathematics. Report No 09–MA–02. Princeton, NJ: National Assessment of Educational Progress.

NAEP (1979c). *Mathematical applications.* Selected results from the Second Assessment of Mathematics. Report No 09–MA–03. Princeton, NJ: National Assessment of Educational Progress.

Ormerod M. B. with Duckworth D. (1975). *Pupils' attitudes to science. A review of research.* Slough: NFER Publishing Company.

Ormerod M. B. and Wood C. (1983). 'A comparative study of three methods of measuring the attitudes to science of 10 to 11 year old pupils'. *European Journal of Science Education, 5,* 77–86.

Pratt J., Bloomfield J. and Seale C. (1984). *Option choice. A question of equal opportunity.* Slough: NFER-Nelson.

Rowell J. A. (1984). 'Towards controlling variables: a theoretical appraisal and a reachable result'. *European Journal of Science Education, 6* 115–30.

Royal Society and Institute of Physics (1982). *Girls and physics.* London: Royal Society and Institute of Physics.

Russell T., Black P., Harlen W., Johnson S. and Palacio D. (1988). *Science at age 11: a review of APU survey findings 1980–84.* London: HMSO.

Ryrie A. C., Furst A. and Lauder M. (1979). *Choices and chances.* Edinburgh: The Scottish Council for Research in Education.

Schofield B., Murphy P., Black P. and Johnson S. (1982). *Science in schools. Age 13: Report No 1.* London: HMSO.

Schofield B., Black P., Head J. and Murphy P. (1984). *Science in schools. Age 13: Report No 2.* London: Assessment of Performance Unit.

Schofield B., Black P., Khaligh N., Murphy P. and Orgee T. (1986). *Science in schools. Age 13: Report No 4.* London: Assessment of Performance Unit.

Schofield B., Black P., Bell J. F. Johnson S. and Murphy P. (1988). *Science at age 13: a review of APU survey findings 1980–84.* London: Assessment of Performance Unit.

Senior R. (1983 unpublished). *Pupils' understanding of some aspects of biological interdependence at age 15.* MA(Ed) thesis, University of Leeds.

Smail B. (1983). 'Getting science right for girls'. In *Contributions to the second GASAT Conference.* Oslo: University of Oslo, Institute of Physics.

SMAIL B. (1984). *Girl-friendly science: Avoiding sex bias in the curriculum.* Schools Council Publication. York: Longmans.

SMAIL B. and KELLY A. (1984). 'Sex differences in science and technology among 11 year old school children: II affective domain'. *Research in Science and Technology Education, 2,* 87–106.

TALL G. (1985). 'Changes in science education due to the schools' response to the Great Debate as indicated by examination entries in England'. *School Science Review, 66,* 668–681.

TODHUNTER I. (1873). *The conflict of studies and other essays on subjects connected with education.* London: Macmillan.

WELFORD G., BELL J., DAVEY A., GAMBLE R. and GOTT R. (1986). *Science in schools. Age 15: Report No 4.* London: Assessment of Performance Unit.

WELFORD G., HARLEN W. and SCHOFIELD B. (1985). *Practical testing at ages 11, 13 and 15.* Science Report for Teachers: 6. London: DES.

WOOLNOUGH B. and ALLSOP T. (1985). *Practical work in science.* Cambridge University Press.

Appendix 1

School questionnaire 1984

Age 15 survey 1984

It is an essential part of the Science Monitoring exercise to relate the test to information about the general provision for Science in schools. This questionnaire is designed to gather such information.

Since this is the last age 15 science survey until 1989, we are anxious to collect more information than usual in a single survey. This particular questionnaire is, therefore, rather lengthy, but we trust you will understand that it is only with such information that full use can be made of the survey results.

Thank you for your co-operation.

General information

Please provide the following general school information:

1. (a) What is the age range of your school? [ ]

   (b) Is your school a grammar, an independent, a secondary modern or a comprehensive school? (Please circle the appropriate letter.) G I S C

   (c) Is your school a girls only school, a boys only school or coeducational? (Please circle the appropriate letter.) B G C

   (d) How many pupils are on your school roll? [ ]

   (e) How many form entries does your school have? (ie if your school is five form entry, enter 5). [ ]

   (f) How many pupils are in your Fifth form year group? [ ]

2. If yours is a new school built during the previous three years, or if your school has been involved in any kind of major restructuring (eg a merger, change of type, expansion etc) in this period, then please indicate this by ticking the box opposite. [ ]

3. (a) How many laboratories does your school have? (A laboratory is here defined as a working space where pupils can do small-group and individual practical work with services – gas, water and electricity.) [ ]

   (b) How many *full-time equivalent* science technicians or laboratory assistants work in your science department? [ ]

4. (a) Please indicate the amount (in £) of your school's *general* capitation (all subjects) allowance *in the last financial year*. (In the case of Independent schools, please indicate the amount of any regular income received by the school for teaching resources.) [£]

   (b) Please indicate the total amount *spent* by your **science** department(s) in this period. (Exclude one-off expenditure from special donation/grants and exclude stationery expenses.) [£]

   (c) Please provide an estimate of the percentage of (b) which was spent on books. [%]

   (d) If your science department(s) received any special donations/grants (eg from the PTA) during this same period please indicate the total amount (in £) and the source(s). [£]

Science teaching in your school

5. Please complete the table matching teachers' subject qualifications in science to the courses they teach to 5th year pupils as listed below.

		Number of teachers with qualifications as shown											
		Biology or equivalent (or Zoology, Biochemistry, etc)			Physics/Phys. Sci. (or Engineering Astronomy, etc)			Chemistry/Chem. Sci. (or Chem. Eng. etc)			Other Science qual. (eg Integ. Sci.) Please specify:	Mathematics qualification	No science or maths. qual.
Teaching group		BSc	BEd	CertEd	BSc	BEd	CertEd	BSc	BEd	CertEd			
Biology	'O' level	...	...	......	...	...	......	...	...	......	....	....	....
	CSE	...	...	......	...	...	......	...	...	......	....	....	....
	Mixed O/CSE	...	...	......	...	...	......	...	...	......	....	....	....
	16+												
Physics	'O' level	...	...	......	...	...	......	...	...	......	....	....	....
	CSE	...	...	......	...	...	......	...	...	......	....	....	....
	Mixed O/CSE	...	...	......	...	...	......	...	...	......	....	....	....
	16+												
Chemistry	'O' level	...	...	......	...	...	......	...	...	......	....	....	....
	CSE	...	...	......	...	...	......	...	...	......	....	....	....
	Mixed O/CSE	...	...	......	...	...	......	...	...	......	....	....	....
	16+												
Human Biology	'O' level	...	...	......	...	...	......	...	...	......	....	....	....
	CSE	...	...	......	...	...	......	...	...	......	....	....	....
	Mixed O/CSE	...	...	......	...	...	......	...	...	......	....	....	....
	16+												
General Science	'O' level	...	...	......	...	...	......	...	...	......	....	....	....
	CSE	...	...	......	...	...	......	...	...	......	....	....	....
	Mixed O/CSE	...	...	......	...	...	......	...	...	......	....	....	....
	16+												

6. Please complete the following table for fifth year pupils who will be entered for 'O' level, 16+ or Mode 1 CSE examinations in science?

N.B. For Mode 3 CSE examination, see question 7.

Subject	Level	Total number of pupils	Appropriate number of:	
			Boys	Girls
Biology	O			
	16+			
	CSE			
Human Biology	0			
	16+			
	CSE			
Physics	O			
	16+			
	CSE			
Chemistry	O			
	16+			
	CSE			
General Science	O			
	16+			
	CSE			
Technology (science department based)	O			
	16+			
	CSE			
Electronics	O			
	16+			
	CSE			
Other science (please specify	O			
	16+			
	CSE			

7. (a) *Mode 3 CSE examinations*
Does your school operate any of the following Mode 3 schemes?

Physics	Yes/No	Environmental science/studies	Yes/No
Chemistry	Yes/No	Rural science/studies	Yes/No
Physical Chemistry	Yes/No	Science at work	Yes/No
Electronics	Yes/No		

Other Science or technology schemes based in the science department Yes/No

If yes to other schemes, please give titles and, if possible, indicate any materials upon which they are based.

(b) Does your school provide courses in science that do not lead to an 'O' level, CSE or 16+ examination? Yes/No

If yes, please give very brief details of course title and form of assessment.

8. If you have 11 to 13 year olds in your school, please indicate if any of the following course patterns apply to them. (Please circle the appropriate letter code.)

(a) General Science studied by all pupils in years 1 to 3.

(b) Biology, Chemistry and Physics studied as separate subjects by all pupils in years 1 to 3.

(c) Biology, Chemistry and Physics studied by more able pupils and General Science by less able pupils in years 1 to 3.

(d) General Science studied by all pupils in years 1 and 2, and separate subjects Biology, Chemistry and Physics by all pupils in year 3

(e) General Science studied by all pupils in year 1, and separate subjects Biology, Chemistry and Physics by all pupils in years 2 and 3.

(f) Other course strategy – please specify.

9. Is it school policy that *at least one science subject should be taken* by fifth year pupils who will probably be entered for:

(a) at least two 'O' levels (not necessarily science)? Yes/No

(b) predominantly CSE examinations? Yes/No

Are any pupils encouraged to take three science subjects? Yes/No
If yes, please give brief details.

10. (a) Do you feel unable to extend the range and type of science courses in your school because of

(i) financial constraints? Yes/No
If yes, please explain.

(ii) lack of appropriately qualified teachers? Yes/No
If yes, please explain.

(iii) logistic constraints, such as lack of laboratory space or space in the curriculum Yes/No
If yes, please explain.

(iv) falling rolls Yes/No
If yes, please explain.

(b) Do you feel unable to increase the number of pupils studying the science courses you currently offer because of:

(i) financial constraints? Yes/No
If yes, please explain.

(ii) lack of appropriately qualified teachers? Yes/No
If yes, please explain.

(iii) logistic constraints, such as lack of laboratory or curriculum space Yes/No
If yes, please explain.

(c) Are you having difficulty in maintaining the level of science provision in your school because of (i) financial constraints? Yes/No
(ii) falling rolls? Yes/No

11. **Provision of text books**

Please complete the table using the code shown for the year groups in your school.

		Biology	Physics	Chemistry	GSc/ CombSci
5th	'O'	......	......	....	....
	16+	......	......	....	....
	CSE				
4th	'O'	......	......	....	....
	16+	......	......	....	....
	CSE				
3rd		......	......	....	....
		......	......	....	....

Code

A each pupil has his/her own book to keep for an extended period of one year or more

B one book is available for each pupil to use for classwork and homework on necessary occasions

C one book is available for each pupil for classwork on necessary occasions

D two pupils have to work simultaneously from one book

E more than two pupils have to work simultaneously from one book

F no text book

G other – please specify

Comments on provision of text books:

If you would care to offer any additional comment which you feel is relevant to this exercise, then please do so here:

Appendix

2

Pupil questionnaires 1984

Two versions of the pupil questionnaire were administered and the first two common pages of the questionnaire are shown here as pages A2.2 and A2.3 (pp137–8).

Pages A2.4 and A2.5 (pp139–40) show the following two pages of one version and pages A2.6 and A2.7 (pp141–2) the other version.

Pupil questionnaire

(1984 APU Science Survey)

By the time you have this questionnaire in front of you, you will most probably have taken one or more of our science tests. We hope it wasn't too bad an experience, and that you'll help us a little more by answering the questions in this booklet. You'll see that we're mainly interested in finding out what you *think* about a number of things – all of which are important. It shouldn't take too long to complete the questionnaire, and we thank you very much for your cooperation.

Pupil number:

(*Please circle*) Boy/Girl

Date of birth ___/___/______

Please tick the relevant boxes to indicate which of these subjects you are studying this year, and then answer the questions about each of them.

		Why did you choose to study this subject?				Do you enjoy studying it? (tick for 'yes')	Please give your reasons							
		It's interesting	*Useful for jobs*	*It's easy*	*Other reason*		*It's interesting*	*Useful for jobs*	*It's easy*	*It's boring*	*It's not important*	*It's difficult*	*Too much homework*	*Other reason*
☐	English	☐	☐	☐	☐	☐	☐	☐	☐	☐	☐	☐	☐	☐
☐	French	☐	☐	☐	☐	☐	☐	☐	☐	☐	☐	☐	☐	☐
☐	German	☐	☐	☐	☐	☐	☐	☐	☐	☐	☐	☐	☐	☐
☐	Italian	☐	☐	☐	☐	☐	☐	☐	☐	☐	☐	☐	☐	☐
☐	Spanish	☐	☐	☐	☐	☐	☐	☐	☐	☐	☐	☐	☐	☐
☐	Welsh	☐	☐	☐	☐	☐	☐	☐	☐	☐	☐	☐	☐	☐
☐	Other language (please name)	☐	☐	☐	☐	☐	☐	☐	☐	☐	☐	☐	☐	☐
☐	Biology	☐	☐	☐	☐	☐	☐	☐	☐	☐	☐	☐	☐	☐
☐	Chemistry	☐	☐	☐	☐	☐	☐	☐	☐	☐	☐	☐	☐	☐
☐	Electronics	☐	☐	☐	☐	☐	☐	☐	☐	☐	☐	☐	☐	☐
☐	General Science	☐	☐	☐	☐	☐	☐	☐	☐	☐	☐	☐	☐	☐
☐	Human Biology	☐	☐	☐	☐	☐	☐	☐	☐	☐	☐	☐	☐	☐
☐	Physics	☐	☐	☐	☐	☐	☐	☐	☐	☐	☐	☐	☐	☐
☐	Rural Science	☐	☐	☐	☐	☐	☐	☐	☐	☐	☐	☐	☐	☐
☐	Technology	☐	☐	☐	☐	☐	☐	☐	☐	☐	☐	☐	☐	☐
☐	Other science (please name)	☐	☐	☐	☐	☐	☐	☐	☐	☐	☐	☐	☐	☐
☐	Art	☐	☐	☐	☐	☐	☐	☐	☐	☐	☐	☐	☐	☐
☐	Computer Studies	☐	☐	☐	☐	☐	☐	☐	☐	☐	☐	☐	☐	☐
☐	Domestic Science	☐	☐	☐	☐	☐	☐	☐	☐	☐	☐	☐	☐	☐
☐	Economics	☐	☐	☐	☐	☐	☐	☐	☐	☐	☐	☐	☐	☐
☐	Geography	☐	☐	☐	☐	☐	☐	☐	☐	☐	☐	☐	☐	☐
☐	History	☐	☐	☐	☐	☐	☐	☐	☐	☐	☐	☐	☐	☐
☐	Mathematics	☐	☐	☐	☐	☐	☐	☐	☐	☐	☐	☐	☐	☐
☐	Other subject (please name)	☐	☐	☐	☐	☐	☐	☐	☐	☐	☐	☐	☐	☐

Are there any *science* subjects which you are not studying this year which you would like to be studying now? If so, please name these in the table below and let us know why you're not studying them.

	Subject				
Reason(s) for not studying it	*Advised against*				
	Not good at it				
	Not interested in it				
	Not useful for jobs				
	Other reason				

	How often do you do these things?					**Would you be interested in doing more given the chance?**				
	Very often	*Often*	*Sometimes*	*Once or twice*	*Never*	*Very interested*	*Quite interested*	*Not sure*	*Not very*	*Definitely not*
Go bird spotting										
Mend things or do jobs in the home using drills, screwdrivers or tools										
Watch a science programme on TV – eg Tomorrow's World, The Living Planet										
Watch science fiction programmes – eg Dr Who, Star Trek, Blake's 7										
Grow plants from seeds										
Play with Scalextric, or electric trains, etc										
Collect and study small animals such as caterpillars and insects										
Take equipment and things apart to see inside them – eg a radio, a hairdryer										
Draw plants or animals that have interested you										
Play games like snooker or billiards										
Read information in books about science – eg Science Now										
Read a science fiction book or magazine – eg Dan Dare, 2001: A Space Odyssey										
Design and then make something from construction kits such as Meccano, Lego, Bolt'n Build.										
Collect fossils or interesting stones, shells or bones										
Repair or try to repair a broken electrical gadget – eg record player; rewire a plug.										
Collect or study wild flowers and plants.										
Use a computer to play games										
Use a computer to do things beside playing games.										
Use a paper pattern to sew something to wear										
Use maps to help you when you go walking, cycling, etc										
Use special kits such as chemistry sets or electronic sets										
Use a magnifying glass or a microscope to see things more clearly.										
Build models from kits such as Airfix.										
Borrow books from the public library.										

How suitable do you think each of these jobs might be–

	for yourself?					**for women?**					**for men?**				
	Very suitable	*Quite suitable*	*Not sure*	*Not really suitable*	*Totally unsuitable*	*Very suitable*	*Quite suitable*	*Not sure*	*Not really suitable*	*Totally unsuitable*	*Very suitable*	*Quite suitable*	*Not sure*	*Not really suitable*	*Totally unsuitable*
Teacher															
Bank manager															
Hairdresser															
Doctor															
Petrol pump attendant															
Clerical worker															
Plumber															
Machinist															
Computer programmer															
Member of Parliament															
Librarian															
Post office worker															
Office cleaner															
Farmer															
Scientist															
Caretaker															
Cashier															
Member of the Forces															
Factory worker															
Bank clerk															
Engineer															
Porter															
Shop assistant															
Solicitor															
Laboratory technician															
Architect															
Secretary															
Chef															
Bricklayer															
Journalist															
Nurse															
Police officer															
Social worker															
University lecturer															
Steelworker															
Garage mechanic															
Professor															
Café worker															
Electrician															
Typist															
Farm worker															
Factory manager															
Driver															
Shopkeeper															

If you have decided yet, which one of these jobs do you hope to have when you leave school/college?...

Many important applications of science and technology are being discussed regularly these days in the newspapers and on television and radio. We are very interested to find out what you know and think about some of these, and would be glad if you'd answer the questions in the table below.

	Have you heard of this?	Does it interest you?	Do you think it's a good thing for society, a bad thing, or can't you say?	Please give your reasons:
Space exploration				
Heart transplants				
'Acid rain'				
Laser technology				
Cable television				
'Test-tube babies'				
Satellite communication				
Nuclear power				
Fibre optics				
Factory farming				
Machine translation				
Intensive agriculture				
Cancer research				
Robotics				
Genetic engineering				
Nuclear weapons				

	How suitable do you think this job would be for you?					How important do you think Biology is for this job?					How important do you think Physics is for this job?				
	Very suitable	*Quite suitable*	*Not sure*	*Not really suitable*	*Totally unsuitable*	*Very important*	*Quite important*	*Not sure*	*Not really important*	*Totally unimportant*	*Very important*	*Quite important*	*Not sure*	*Not really important*	*Totally unimportant*
Teacher															
Bank manager															
Hairdresser															
Doctor															
Petrol pump attendant															
Clerical worker															
Plumber															
Machinist															
Computer programmer															
Member of Parliament															
Librarian															
Post office worker															
Office cleaner															
Farmer															
Scientist															
Caretaker															
Cashier															
Member of the Forces															
Factory worker															
Bank clerk															
Engineer															
Porter															
Shop assistant															
Solicitor															
Laboratory technician															
Architect															
Secretary															
Chef															
Bricklayer															
Journalist															
Nurse															
Police officer															
Social worker															
University lecturer															
Steelworker															
Garage mechanic															
Professor															
Café worker															
Electrician															
Typist															
Farm worker															
Factory manager															
Driver															
Shopkeeper															

If you have decided yet, which one of these jobs do you hope to have when you leave school/college? ....

	Have you studied...? (tick for 'yes')	Would you like to know more about it? *Definitely yes*	*Quite a lot*	*Not sure*	*Not very*	*Definitely not*
What gives fireworks different colours?						
How plants and animals respire.						
How to see round corners.						
How substances dissolve.						
Can vegetarians (people who don't eat meat) live without other animals?						
How does friction work?						
Why it's dangerous to throw aerosol cans on a bonfire.						
How animals respond to their environment.						
Is there such a thing as perpetual motion?						
How acids and bases react.						
How can we stop insects becoming resistant to insecticide?						
What is necessary for a steady current to flow between the terminals of a battery?						
Which materials are best for making saucepans?						
How animals reproduce.						
How to lift heavy weights with bags of air.						
How a change of state is caused.						
How does cutting down forests affect the air we breathe?.						
How mass differs from weight.						
Why jewellery is made from silver and not aluminium.						
How animals are adapted to the kind of food they eat.						
How bald tyres increase the chance of car accidents.						
Properties of metals.						
How to grow plants without soil.						
How to find the average speed of an object.						
Where does the water on the outside of cold drink cans come from?						
How plants and animals depend on each other.						
How balloons can be used to help teach deaf children.						
How elements vary in their relative activity.						
How do our lungs work?						
How pressure depends on force and area.						
Why is the sea salty?						
How animals are structurally adapted to their environment.						
How to mend a torch which doesn't work.						
How metals are distinguished from non-metals.						
Why are some twins identical?						
How light travels.						
How it's good that toothpaste is a base but bad that apples are acidic.						
Our ears and how we hear.						
Why it's hard to move in a space-craft.						
What the difference is between a solid, a liquid and a gas.						
Why the kiwi bird might become extinct.						
How energy is transferred from one thing to another.						
Why fibre glass and not metal is used for making cars that last.						
How plants make food.						
How do you prove you've broken a world speed record?						
How animals are trained.						
How sound travels.						
How to cross-breed goats and sheep.						

Appendix

3

Some detailed pupil questionnaire results 1984

Table A3.1
Responses to the enquiry about the suitability of different jobs for women and for men
(Percentage of pupils giving indicated response* – 6,200 pupils in total)

	for women					for men				
	VS	QS	U	NS	TU	VS	QS	NU	NS	TU
Teacher	42	45	9	2	1	46	44	8	2	1
Bank manager	16	32	24	21	8	69	26	3	1	1
Hairdresser	80	16	2	1	1	20	38	16	15	10
Doctor	37	43	13	5	3	69	26	3	1	1
Petrol pump attendant	7	15	19	29	29	42	39	12	4	2
Clerical worker	35	34	20	5	5	23	36	27	9	5
Plumber	5	9	16	31	40	67	27	4	1	1
Machinist	17	22	23	18	19	46	25	15	7	7
Computer programmer	30	43	19	5	4	52	39	7	1	1
Member of Parliament	33	36	17	7	8	55	31	9	2	3
Librarian	59	32	6	2	1	18	31	24	18	9
Post office worker	38	40	15	4	2	30	41	19	7	3
Office cleaner	43	37	11	5	4	10	15	21	26	27
Farmer	9	18	23	25	25	70	23	5	1	2
Scientist	30	38	19	7	6	60	32	6	1	1
Caretaker	11	18	24	25	22	51	34	10	3	2
Cashier	49	38	9	2	2	27	33	23	12	6
Member of the Forces	24	32	20	13	12	72	20	5	1	2
Factory worker	24	35	21	12	7	41	36	15	5	3
Bank clerk	34	41	16	5	4	47	39	10	2	1
Engineer	9	17	24	26	15	71	23	4	1	1
Porter	6	10	22	28	34	49	35	11	3	2
Shop assistant	54	36	7	2	2	22	35	24	14	5
Solicitor	28	34	22	9	6	61	29	7	1	1
Laboratory technician	29	35	21	8	7	43	36	15	4	2
Architect	21	27	28	14	10	55	31	10	2	2
Secretary	78	15	4	2	2	12	18	22	25	23
Chef	42	31	14	6	7	63	27	7	2	2
Bricklayer	5	6	13	23	54	72	21	4	1	1
Journalist	41	37	14	4	4	47	37	12	2	2
Nurse	80	14	3	1	1	21	29	20	13	17
Police officer	31	36	17	8	7	71	20	5	1	2
Social worker	54	29	12	3	3	28	32	24	10	6
University lecturer	29	32	24	8	7	51	32	13	2	2
Steelworker	4	5	14	23	55	65	26	6	1	1
Garage mechanic	6	11	17	24	42	71	23	4	1	1
Professor	21	27	24	12	16	57	29	10	2	2
Café worker	44	38	12	4	3	17	23	27	21	12
Electrician	10	16	25	23	25	64	27	5	2	2
Typist	77	15	4	1	2	10	16	23	25	26
Farm worker	9	17	25	23	25	60	29	7	2	2
Factory manager	17	24	27	17	15	62	29	7	1	1
Driver	22	24	22	17	14	67	26	5	1	1
Shopkeeper	54	33	9	2	1	42	33	15	7	4

* VS–Very Suitable, QS–Quite Suitable, U–Unsure, NS–Not Suitable, TU–Totally Unsuitable.

Table A3.2

Responses to the enquiry about the suitability of different jobs for themselves

(Percentage of boys and girls giving indicated response* – 5,400 pupils of each sex)

	boys					girls				
	VS	QS	U	NS	TU	VS	QS	NU	NS	TU
Teacher	3	13	20	29	35	7	21	21	30	21
Bank manager	8	19	23	23	27	4	15	21	30	30
Hairdresser	1	4	9	19	66	17	27	19	19	18
Doctor	6	13	18	24	39	5	13	18	26	38
Petrol pump attendant	4	13	19	22	41	1	5	10	20	64
Clerical worker	3	12	24	23	39	9	22	22	20	27
Plumber	8	24	23	20	25	<1	3	7	18	71
Machinist	9	19	24	19	29	2	8	14	21	55
Computer programmer	15	20	19	17	29	4	12	19	21	44
Member of Parliament	7	9	12	13	58	4	6	10	13	68
Librarian	2	8	15	22	53	6	23	20	22	30
Post office worker	3	14	23	24	36	5	20	25	24	26
Office cleaner	1	3	8	17	71	2	7	11	22	58
Farmer	12	17	17	16	38	5	11	12	17	55
Scientist	9	16	19	18	38	4	9	14	17	56
Caretaker	3	9	16	23	50	1	3	9	21	67
Cashier	6	16	22	22	34	10	30	22	18	21
Member of the Forces	22	23	19	11	25	12	20	20	15	33
Factory worker	6	16	21	21	36	3	10	15	20	51
Bank clerk	8	21	21	19	31	10	24	22	18	26
Engineer	23	29	20	12	16	3	7	13	20	58
Porter	4	9	19	24	44	1	3	9	21	66
Shop assistant	6	18	22	22	32	16	35	19	15	15
Solicitor	9	17	20	16	38	8	16	21	19	37
Laboratory technician	5	15	20	20	39	4	11	14	18	54
Architect	12	21	20	16	32	5	12	14	19	50
Secretary	2	5	13	22	59	23	29	17	14	17
Chef	12	19	21	16	31	13	22	19	17	29
Bricklayer	16	20	18	16	30	1	3	6	14	75
Journalist	9	18	20	18	34	11	20	21	16	32
Nurse	2	4	9	15	70	16	19	17	19	29
Police officer	12	20	19	14	34	10	16	19	17	38
Social worker	3	7	17	19	54	14	22	20	15	28
University lecturer	3	7	13	15	61	2	5	11	16	67
Steelworker	9	15	19	19	39	<1	1	4	11	84
Garage mechanic	21	21	18	15	25	2	4	7	13	73
Professor	6	10	14	16	55	2	4	10	14	69
Café worker	4	8	15	22	52	7	18	20	20	35
Electrician	21	27	20	13	19	2	6	12	18	61
Typist	2	4	10	18	67	23	25	17	13	22
Farm worker	12	15	16	15	41	5	10	13	16	55
Factory manager	11	22	25	15	26	4	10	17	20	49
Driver	30	28	18	10	14	11	17	20	16	36
Shopkeeper	11	18	24	18	31	17	28	22	15	18

* VS–Very Suitable, QS–Quite Suitable, U–Unsure, NS–Not Suitable, TU–Totally Unsuitable.

Table A3.3

Perceptions of the importance of Biology and Physics to various occupations

(Percentage of pupils giving indicated response* – 6,100 pupils)

	Biology					Physics				
	VI	QI	U	NV	NI	VI	QI	U	NV	NI
Teacher	15	32	31	16	6	18	31	32	13	6
Bank manager	1	3	14	35	47	3	12	23	29	33
Hairdresser	11	22	16	19	33	4	10	18	23	45
Doctor	85	8	4	1	3	46	28	14	6	7
Petrol pump attendant	1	2	11	16	71	3	10	17	19	50
Clerical worker	3	6	27	25	40	3	8	29	24	36
Plumber	1	4	19	23	53	17	22	22	14	25
Machinist	1	4	19	25	48	17	22	22	14	25
Computer programmer	2	6	19	25	48	25	27	20	12	16
Member of Parliament	3	7	19	19	52	5	9	23	19	45
Librarian	2	9	19	23	47	2	8	20	24	47
Post office worker	1	2	13	25	59	1	4	16	26	53
Office cleaner	1	3	9	14	72	1	2	10	15	72
Farmer	32	35	14	6	12	12	26	25	14	24
Scientist	80	11	4	1	4	83	7	4	1	4
Caretaker	2	3	14	9	62	3	6	16	18	58
Cashier	1	3	14	22	60	2	6	17	23	53
Member of the Forces	5	19	30	16	29	9	22	29	14	26
Factory worker	1	4	18	21	56	3	12	22	19	43
Bank clerk	1	5	18	23	52	4	12	22	22	40
Engineer	3	9	22	22	44	41	23	14	7	15
Porter	1	2	13	18	67	2	3	14	19	63
Shop assistant	1	3	14	21	61	1	3	14	21	61
Solicitor	4	8	22	19	46	6	11	25	18	40
Laboratory technician	52	23	11	4	10	60	21	9	2	7
Architect	5	10	25	20	40	21	20	24	12	23
Secretary	1	4	17	24	55	2	4	19	24	51
Chef	14	25	20	13	27	3	8	21	22	46
Bricklayer	1	3	12	18	66	4	13	17	18	47
Journalist	3	7	20	21	48	3	7	22	20	47
Nurse	72	14	6	2	6	29	27	21	9	15
Police officer	7	17	29	16	30	6	15	30	18	31
Social worker	6	16	28	18	33	3	8	28	21	40
University lecturer	27	21	27	7	17	30	21	27	6	15
Steelworker	2	5	20	19	55	14	23	24	13	26
Garage mechanic	3	5	17	20	56	25	26	19	9	21
Professor	46	18	20	4	12	50	17	19	4	10
Café worker	2	6	15	17	59	3	3	13	17	64
Electrician	3	8	22	21	47	47	22	12	6	13
Typist	1	3	14	20	62	2	4	17	21	56
Farm worker	22	31	18	9	20	7	19	25	16	34
Factory manager	3	7	24	20	47	7	16	27	17	34
Driver	1	2	15	20	61	5	11	19	19	46
Shopkeeper	1	5	17	19	57	1	3	16	20	59

* VI–Very Important, QI–Quite Important, U–Unsure, NV–Not Very Important, NI–Not at all Important.

Appendix

4

Notes on sampling, test distribution, marking and survey analysis

The pupils who took part in the 1984 survey were selected, as usual, according to a two-stage stratified cluster sampling scheme. In the first stage a random sample of schools was selected, and this was followed in the second stage by the random selection of pupils of the appropriate age from within each of the participating schools. The sample survey scheme and other aspects of this appendix are described in the technical review of the science survey programme (Johnson, 1988).

Selection of schools

Before selection began, the school population was stratified with respect to the variables *size, type* and *region*. The relevant regional classification within England is shown in Table A4.1.

Four *size of age group* classifications were imposed: *up to 80 pupils, 81–160 pupils, 161–240 pupils* and *more than 240 pupils.*

Four *type of school* categories applied in England and Wales: comprehensives with pupils up to age 16, comprehensives with pupils up to age 18, other maintained schools and independent schools. In Northern Ireland, technical colleges were distinguished from schools, and schools were identified as grammar or intermediate, and further subdivided according to their management system into controlled or maintained/voluntary schools.

Table A4.1 *The regions of England*

North	Midlands	South
Merseyside*	West Midlands*	Greater London*
Greater Manchester*	Hereford and Worcester	Bedfordshire
South Yorkshire*	Shropshire	Berkshire
West Yorkshire*	Staffordshire	Buckinghamshire
Tyne and Wear*	Warwickshire	East Sussex
Cleveland	Derbyshire	Essex
Cumbria	Leicestershire	Hampshire
Durham	Lincolnshire	Hertfordshire
Humberside	Northamptonshire	Isle of Wight
Lancashire	Nottinghamshire	Kent
North Yorkshire	Cambridgeshire	Oxfordshire
Northumberland	Norfolk	Surrey
Cheshire	Suffolk	West Sussex
		Isle of Scilly
		Avon
		Cornwall
		Devon
		Dorset
		Gloucestershire
		Somerset
		Wiltshire

* Metropolitan counties in 1984.

Within England schools were selected from within each region-by-type-by-size classification in numbers which reflected their presence in the school population as a whole (ie schools were selected by proportional random sampling): in other words, if x per cent of the schools in England containing 15 year old pupils were 11–16 schools of size 81–160 pupils in the Midlands, then x per cent of the English school sample should be of this type. Within Wales and Northern Ireland schools were also selected by proportional random sampling from the type-by-size or management-by-size classifications, respectively, but these countries were deliberately over-represented relative to England in the final sample so that pupil performance estimates of reasonable accuracy could be produced.

Table A4.2 provides details of the numbers of schools invited to take part in the survey, and of those which finally participated. The participation rates for England, Wales and Northern Ireland were 82 per cent, 73 per cent and 68 per cent respectively.

Table A4.2 *The sample of schools*

	England	Wales	Northern Ireland
Invited to take part	331	144	145
Unable to take part	55	35	42
No reply	1	1	2
Initial acceptance, later decline	3	1	3
Tests not returned or returned unused	—	2	2
Schools participating	272	105	99

Selection of pupils

The pupils chosen to take part in the survey were selected from all of those in the participating schools who were born between 1st September 1968 and 31st August 1969. Pupils were selected by reference to their dates of birth, and the range of birth dates specified varied according to the size of age-group in the school, so that roughly equal samples of pupils would be selected from each school.

The circus practical tests, for economic reasons, were administered to groups of nine pupils at a time (pupils working independently). Pupils were randomly selected for the practical tests from those taking the written tests. All schools took two circus practical tests.

The only pupils explicitly excluded from the survey were those in special schools or in units designated as 'special' within normal schools. However, the Headteacher of each selected school was told that discretion could be used in withdrawing particular pupils from the testing sessions if it was felt that participation would cause undue distress. About 14,000 pupils were chosen to take part in the survey, and just 56 pupils were withdrawn from testing by their Headteachers. Clerical errors of one kind or another resulted in the overall loss of test results for fewer than one per cent of pupils at the final analysis stage.

Test administration

The questions which were used in this survey to represent each subcategory were chosen at random from the pool of questions for that subcategory (ie a 'domain-sampling' approach to question selection was employed). Table A4.3 shows the number of questions selected from each subcategory pool. This complete set of questions was then subdivided into three, four or six subtests to be administered to different, but similarly representative, random samples of pupils.

Table A4.3 *The sample of questions*

Subcategory	Number of subtests	Number of questions selected
Using graphs, tables and charts	6	90
Making and interpreting observations	3	45
Interpreting presented information	6	78
Applying biology concepts	6	66
Applying physics concepts	6	66
Applying chemistry concepts	6	66
Planning parts of investigations	4	60

Twenty-two different written test packages (each in two versions both containing the same questions but one version presenting these in the reverse order to the other) were administered to pupils in this survey. Most test packages consisted of questions from two different subtests. Each package was intended to last about an hour. There were four practical circuses; three circuses containing questions drawn from the subcategory pool **Making and interpreting observations** and one circus comprised of the fixed test of **Use of apparatus and measuring instruments.** In addition, six individual practical investigations were administered to some of the sample pupils. There were also several packages which were not samples of questions from the subcategory pools but questions designed to investigate various aspects of science performance.

Each survey school was given a variety of the written test packages to be distributed at random among its sample pupils. Each pupil took just one of these. In addition, in each of a random subsample of the participating schools, subsamples of pupils taking written tests took part in a practical circus. Testing took place during November in the three countries.

Pupils who were in the sample but who were absent on the day on which the school undertook a written test session were given the relevant test if they attended school at any time within two weeks of the school's main test session. This was an attempt to avoid introducing unnecessary bias into the pupil sample which would occur if, for example, persistent absentees happened also to be the lower performers in general. 633 pupils did not complete their written tests because of absence from school during the two week period.

Marking

The testers who were trained to administer the individual or group practical tests were also trained to mark the results of this testing. A detailed checklist has been devised for each individual category 6 practical task, and the administrators simply coded these checklists as the pupil attempted the tasks. In the group practical circuses, an occasional question would be marked at the time of testing but most would be marked as written responses at the end of the day.

For the written tests, pairs of markers were trained to mark one or other of the test packages, one of the pair marked all of the 'A' version scripts the other marked all of the 'B' version scripts (these represent equivalent random samples of between 250 and 300 scripts).

Analysis

The school questionnaire data were analysed with the aid of SAS (Helwig and Council, 1979). As the school sample was produced by a simple random sampling procedure within each stratum, it would be appropriate to apply the usual chi-square and other significant tests to the data if the statistical significance of any of the subgroup differences is of interest.

Throughout the report, any results presented for individual questions are raw sample statistics and have not been weighted in any way.

Population mean performance stimates and their associated estimated variances were produced test package by test package for the subcategories contained in them. Where a subcategory was represented in two or more different test packages, the separate, independent performance estimates produced for these were combined.

The procedure requires that an appropriate subcategory percentage 'subtest score' be computed for each individual pupil, this being the simple sum of the pupil's scores on relevant questions as a percentage of the total possible score (after an initial adjustment of all question scores onto a common mark scale).

When producing the population estimates for each science subcategory in each test package it was necessary to weight the raw sample data to take account of the complex sampling scheme which was used to select the pupil sample. The method adopted here to produce a population mean estimate for any particular subgroup ('North', 'Comprehensive to 16' etc) was first to multiply each school mean by that school's size (ie by the number of pupils in that school of the appropriate age), to sum the resulting figure over all schools in that subgroup sample, and then to divide the whole by the total number of pupils of the appropriate age in those sample schools (technically termed a biased ratio estimate). Estimates were produced in this way for sample breakdowns by the stratifying variables, ie school location, school type and region. Weighted variance estimates were produced simultaneously – the relevant formula is to be found in any text on survey sampling.

The overall population mean and variance estimates, and those for boys and girls separately, were produced by appropriately weighting the separate regional estimates before combining.

The subcategory mean estimates reported in Chapter 1 were produced by averaging the separate estimates produced for each test package which contained questions from that subcategory. The standard errors associated with these final mean scores were produced in the usual way for an average of independent variates by dividing the square root of the sum of the separate variance estimates by n (where n is the number of test packages).

It should be noted that these standard errors, though following conventional practice applying recognised formulae, take account only of estimation errors arising from the sampling of pupils and schools. They do not allow for estimation errors arising from the sampling of questions, nor those arising from interactions between pupils or schools and questions. Computation of standard errors which should take this contribution into account is complex and time-consuming; the results of some preliminary analysis of the survey data do, though, suggest that the standard errors quoted in Chapter 1 could be multiplied by a factor of about three to approximate those which would apply if the question sampling influence were in fact taken into account (see Johnson and Bell, 1985; Johnson, 1988).

Appendix

5

Use of apparatus and measuring instruments: 1984 fixed-test results

The fixed test in 1984 was administered at both ages 13 and 15. It was adapted from that used at age 15 in 1982 and incorporated modifications to make it an appropriate test at both ages. Each station of the circus was timed to be completed in eight minutes, and each pupil attempted the questions at nine circus stations.

The results for the test at age 15 are presented below with little descriptive prose. Discussion of these results is incorporated within the main review of the category in Chapter 6.

In 1984 for the first time multiple-part questions were presented in two orders. This was to reduce the effect of pupils running out of time and not attempting the latter parts of multiple questions. (The (a) and (b) orders for 'Roundabout' are both reproduced in Section A5.3 to exemplify the method chosen to produce two orders for the multiple part questions.) Administration of any ordering sequences of greater complexity would have proved too cumbersome both for printing and data collection. In the event two orders proved sufficient to eliminate the effects mentioned above. In 1982 the percentage of pupils not recording a reading for 'Instruments' rose steadily over the final four parts of the question (Gott *et al*, 1985 pp27–28); a trend which is not present in the equivalent question in 1984 – see Table A5.1 (p150).

Comparison of performance on the two orders within each of the questions so ordered has showed no difference between pupils regardless of internal question order; in other words, meeting one instrument before another or attempting one limited task before another confers neither advantage nor disadvantage to candidates.

In 1982 all pupil values were collected and used to present the distribution of their responses around the set or expected values. This was repeated in 1984, and the patterns of response are noteworthy for the marked similarity to the 1982 results. For that reason detailed histograms relating to the 1984 data are not given in this appendix.

In addition to pupil values in 1984 the practical supervisors were asked to keep a record of all set values throughout the running of a circus. The purpose of this was twofold, apart from giving the supervisors a constant check for their own administrative uses. Firstly the intention was to plot any change or slippage in a value over the duration of the circus. For instance, a vacuum flask of water set at 43°C might be expected to lose heat so that the pupils coming to this station after an hour might encounter a reading of 42°C or even 41°C. Knowing the exact value encountered by each pupil has enabled a more accurate scoring to be attempted (by awarding marks for deviation of pupil value from encountered value rather than from a specified value which may have slipped) as well as making it probable that the practical supervisor would be able to adjust the temperature so that 43°C was available to all pupils. The second purpose was to inform potential users of these types of questions administered in in-school testing of practical skills as to the nature and extent of problems encountered by the APU in maintaining absolute consistency throughout a practical testing programme.

Comment has been made on this aspect of the survey in Section 6.2 of Chapter 6.

A5.1 Reading scales on measuring instruments

'Scale readings' was a question very similar to 'Instruments' used in 1982 (Gott *et al*, 1985, p27) and is not reproduced here. A selection of measuring instruments was arranged at a single circus station with each instrument pre-set at a specific value.

A5.2 Using measuring instruments

Three types of question were involved:

(1) Measuring out set quantities of materials to be checked by the supervisor.

'Deliver', the question used to test pupils' ability in this sphere, was identical to that employed in the 1982 survey at age 13. Table A5.2 (p150) indicates the tasks and gives the proportion of pupils measuring out quantities within certain ranges of the value required.

Table A5.1 *'Scale readings' – performance of pupils reading various pre-set measuring instruments**

Instruments	Set value	Calibration: Marked divisions	Calibration: Sub-divisions	Range of tolerance	% Pupils (n = 612): Within that range	% Pupils (n = 612): % Pupils not responding
Measuring cylinder	42 cm^3	10 cm^3	×1	41–43	51	1
Forcemeter	16 N	10 N	×2	14–18 15–17	80 66	3
Manometer	18 cm H_2O	1 cm H_2O	×1	17–19	22	12
Thermometer	43°C	10°C	×1	42–44	85	2
Stopclock	437 S	5 S	×1	436–438	51	5
Ammeter	0.24 A	0.1 A	×0.02	0.22–0.26 0.23–0.25	35 17	2
Voltmeter	1.1 V	1 V	×0.1	1.0–1.2	84	3
Lever arm balance	117 g	50 g	×2	115–119 116–118	40 34	6
Ruler	303 mm	1 cm	×1 mm	302–304	54	1

* Where minor scale graduations are in 'twos', then two ranges of tolerance (and percentage of pupils) are presented – the first to encompass the proportion of pupils within one scale gradation of the set value, and the second giving the percentage of pupils within one unit (half of the minor scale graduation) of the set value.

Table A5.2 *'Deliver' – performance of pupils measuring out fixed quantities of materials*

Instrument	Minor scale division	Quantity required	Range of tolerance	% Pupils (n = 504): Within that range	% Pupils (n = 504): Not responding
Rule	0.1 cm	47.3 cm papertape	47.2–47.4 47.1–47.5	63 86	3
Measuring cylinder	1 cm^3	55 cm^3 water	54–56 53–57	60 76	11
Lever arm balance	2 g	68.0 g plasticine	66–70 64–72	52 68	1
Lever arm balance	2 g	82.0 g sand	80–84 78–86	22 32	7

Table A5.3 *'Measuring' – performance of pupils using instruments to measure various quantities*

Instrument	Minor scale division	Quantity measured	Range of tolerance	% Pupils (n = 515): Within that range	% Pupils (n = 515): Not responding
Rule	0.1 cm	Extension of spring under load (4.0 cm)	3.9–4.1 3.8–4.2	32 39	2
Forcemeter	0.1 N	Force to lift a given mass (5.5 N)	5.4–5.6 5.3–5.7	69 75	2
Measuring cylinder	1 cm^3	Capacity of a beaker (136 cm^3)	135–137 134–138	11 16	4
Stopclock	1 second	Duration of a light flash (4.0 S)	3.5–4.5 3.0–5.0	71 94	2

(2) Using measuring instruments to measure various quantities.

The question 'Measuring' used here was very similar to those in use in 1982 at each of the ages 13 and 15. Table A5.3 shows the tasks and gives the proportion of pupils making their measurements within certain ranges of accuracy. In the case of the rule two readings and a subtraction were necessary, and the stopclock had to be both started and stopped.

(3) Use of measuring instruments in limited tasks involving standardized procedures.

In 'Techcheck' pupils were required to carry out three tasks and a selection of their behaviours was recorded on a checklist by the participating school's teacher. The percentages of pupils performing certain actions in the three tasks are recorded in Table A5.4 (p151).

Table A5.4 *'Techcheck' – performance of pupils on various aspects of three tasks requiring the use of instruments in standard techniques*

Task 1 Measure the temperature of water (c60°C) using a thermometer (n = 513)

Action	*% Pupils*	*Action*	*% Pupils*
Used thermometer out of case	75	Used thermometer still in case	24
Read thermometer bulb down	96	Read thermometer upside down	1
Read thermometer, bulb immersed	79	Read thermometer, bulb withdrawn	17
Waited for final temperature	86	Read thermometer immediately upon immersion	9

Task 2 Determine the average volume of a marble given several similar marbles (n = 513)

(a) *Main method employed in experiment*	*% Pupils*
Used a measuring cylinder – mostly by water displacement	49
Used spring balance and assumed S.G of 1	40
Measured diameter and calculated using recalled formula for volume	6
No visible attempt	5

(b) *Actions of pupils using measuring cylinder method (n = 250)*

Action	*% Pupils*	*Action*	*% Pupils*
Used water in cylinder	93	Used cylinder without water	7
Added water **before** adding marbles	88	Added water **after** adding marbles	5
Read water level before **and** after adding marbles	85	Read level **only after** marbles added	3
Read cylinder in upright position	88	Read tilted cylinder	5
Read cylinder with eye **at** water level	72	Read cylinder with eye not at water level	21

(c) *Number of marbles used – all three methods (n = 513)*

	% Pupils
Used one marble only	61
Used more than 1 marble	34

Task 3 Determine the period of a pendulum swing (n = 513)

Action	*% Pupils*	*Action*	*% Pupils*
Swing initiated with bob <10 cm from vertical	76	Bob pulled back >10 cm from vertical	19
Pendulum bob released	92	Swing bob into suspensory apparatus	13
Timed more than one swing	30	Bob pushed	3
Stopclock started		Timed one swing only	65
–bob at centre of swing	33	No attempt to do experiment	5
–bob at extreme(s)	58		
–arbitrarily	5		
Clock stopped in order to take final reading	83		

Table A5.5 *Estimating. Distribution of pupil responses for 'Estimating' and 'Guessit', two very similar questions used in 1984 and 1982*

	1984 'Estimating' % Pupils (n = 508)					1982 'Guessit' % Pupils (n = 754)				
		Within % of value					Within % of value			
Physical quantity	Value presented	10%	20%	50%	Tendency to over/ underestimate	Value presented	10%	20%	50%	Tendency to over/ underestimate
Volume	300 cm^3	6	12	29	under	126 cm^3	4	17	33	under
Volume	25 cm^3	6	20	33	under	6 cm^3	5	21	32	neither
Area	200 cm^2	6	10	30	under					
Area	3 cm^2	17	21	53	over	15 cm^2	8	30	39	under
Length	30 cm (rod)	49	69	85	under	37 cm (rod)	42	67	84	under
Length	30 cm (circle of wire)	23	38	76	under					
Mass	5,000 g	5	11	18	under	5 kg	13	21	24	neither
Mass	20 g	9	16	30	under	75 g	4	7	27	under
Temperature	30°C	10	20	50	under	40°C	21	41	46	under
Force	20 N	10	15	36	neither	40 N	6	21	25	under
Time	20 s	28	45	78	neither	3.5 s	36	37	83	over

The teacher-observers also recorded the incidence of repeating the experiment and whether the subsequent experiment(s) resulted in improvements. Very few pupils (c 10%) checked their readings or repeated their efforts either to confirm or to improve on previous attempts. However, pupils may have felt constrained by a lack of time (eight minutes for the whole question comprising three tasks) and so little significance is attached to this facet of performance in this context.

A5.3 Estimating physical quantities

Two distinct types of question were used to test pupils' ability to estimate in the absence of measuring instruments:

(1) 'Guesstimating' the magnitude of physical quantities.

Given a 'variety of quantities' of known value pupils were asked 'How long is...?' How heavy is...?' and required to make their best estimate (guess).

Table A5.5 (p151) shows the results to two multiple part questions of very similar type used in 1984 and in 1982. The 1982 results are included to increase the data upon which are based the findings discussed in Chapter 6.

(2) Delivering a specified amount of material.

The question 'Roundabout' is reproduced below in A and B orders to illustrate the method chosen to produce two orders for multiple-part questions.

As for 'Estimating' the results for 'Roundabout' are shown as percentages of pupils within 10%, 20% and 50% of the specified value, and are given in Table A5.6.

Table A5.6 *'Roundabout'. Distribution of pupil responses for 'Roundabout'*

Task (without using measuring instruments):	Value requested	Pupils (n = 505) within set value 10%	20%	50%
a) cut length of paper tape	50 cm	15	31	56
b) put dried beans into bag	100 g	15	29	62
c) put porridge oats into bag	100 g	22	37	78
d) put water into beaker	100 cm^3	17	33	76
e) draw line	11 cm	13	21	53

A5.4 Using a microscope and hand lens

Pupils were given three slides which carried reduced upper case alphabet letters. They were required to view them and to record what they saw, but were not told that their subject comprised letters. The question was in three parts. The first required use of a hand lens. The second required pupils to focus the low power objective lens of a microscope with the slide in place. The third part demanded that pupils place the slide onto the microscope stage and then search to locate the characters. The proportions of pupils completing aspects of each of the three tasks are recorded in Table A5.7 (p153).

A5.5 Using meters in an electrical circuit

Pupils were presented with a simple circuit already connected. They had to select the correct meter, insert it in an appropriate place in the circuit and record the value (see Gott, 1984). The results for 1984 are shown in Table A5.8 (p153).

A

In this question you will not be allowed to use any measuring instrument. You will have to make as good a guess as you can.

(a) Cut off a length of tape 50 cm long. Put it in the envelope to one side.

(b) Take a plastic bag with your letter on it. Put 100 grams of dried beans in it. Seal the bag and 'post' it in the box provided.

(c) Take the other plastic bag with your letter on it. Put 100 grams of porridge oats in it. Seal it and post it in the same box.

(d) Take the beaker with your letter on it. Put 100 cm^3 of water into it from the tap. Put the beaker and water carefully into the box provided.

(e) Draw a line 11 cm long in the space below. You may use the 'straight edge' to help you.

B

In this question you will not be allowed to use any measuring instrument. You will have to make as good a guess as you can.

(a) Take the beaker with your letter on it. Put 100 cm^3 of water into it from the tap. Put the beaker and water carefully into the box provided.

(b) Draw a line 11 cm long in the space below. You may use the 'straight edge' to help you.

(c) Cut off a length of tape 50 cm long. Put it in the envelope which has your letter on it, and set the envelope to one side.

(d) Take a plastic bag with your letter on it. Put 100 grams of dried beans in it. Seal the bag and 'post' it in the box provided.

(e) Take the other plastic bag with your letter on it. Put 100 grams or porridge oats in it. Seal it and post it in the same box.

Table A5.7 *'Lenses' – percentage of pupils giving various types of response*

	% Pupils (n = 510)
(a) Using a hand lens (×8), view reduced characters and record what was seen	
All characters recorded correctly without sequence or orientation disorders	59
characters recorded with sequence disorder/wrong letter inserted	10
characters recorded with orientation disorders/upside down etc	15
characters recorded but with both sequence and orientation disorders	4
other errors – letters omitted etc	9
no recorded response	3
(b) Focus microscope (low power) on slide, and record what was seen	
All characters recorded correctly	79
sequence disorder/wrong letter inserted	4
orientation disorders	1
both sequence and orientation disorders	<1
other errors – letters omitted etc	11
no recorded response	5
(c) Place slide on stage, locate characters, focus microscope, view and record	
All characters recorded correctly	81
sequence disorder/wrong letter inserted	4
orientation disorders	<1
both sequence and orientation disorders	0
other errors – letters omitted etc	5
no recorded response	10

Table A5.8 *'Amvocirc' – proportion of pupils carrying out various actions*

	% Pupils (n = 515)
(a) Selection of ammeter and insertion in circuit	
chose ammeter	67
connected correctly in series (read accurately)	26(18)
connect across switch	5
connect across cell	12
connect across bulb A	10
connect across bulb B	4
connect across both bulbs	3
chose ammeter but no circuit to assess	7
chose voltmeter	25
no response/no circuit attempted	8
(b) Selection of voltmeter and insertion in circuit	
chose voltmeter	80
connected correctly across bulb A (read accurately)	30(26)
connected across bulb B	3
connected across both bulbs	1
connected across switch	2
connected across cell	6
others incorrect insertion (in series)	21
chose voltmeter, but no circuit to assess	17
chose ammeter	15
no response/no circuit attempted	5

Appendix

6

Performing investigations: 'Catalase' and 'Springload'

A6.1 Introduction

These two tasks were selected for use in surveying because they represented a broadly similar logical structure, though with differing contexts and differing levels of complexity in setting up the major variables. The tasks as presented to the pupils are shown in Figures A6.1 and A6.2 (pp155–6). The practical assessors completed a behavioural and a summary checklist for each pupil. For some account of the methodology involved and the analytical methods used for checklists, see earlier reports, especially the second report (Driver *et al*, 1984, Chapter 6). Some of the difficulties associated with this class of tasks are noted briefly here. As usual they stem from the need to reconstruct 'logical', isolated experiments from the pupil's activities, but are exacerbated by the many alternative experimental designs which are available in these cases. Sequences of trials often lend themselves to diverse interpretations. In addition, a pupil may be undertaking a sequence of isolated experiments, may undergo significant learning during the activities, may relate chronologically disparate data and so on. These problems vary in differing aspects of the task. In assessing experimental design the focus has been on the *best* experiment extractable from the sequence of trials. In assessing operationalization of the key dependent variable a more chronological approach has been attempted, especially in regard to shifts between qualitative and quantitative approaches. This has been done independently of the trials used in the experimental design. In the main these judgements have been based on detailed analysis of the assessor's behavioural checklist. A few results have been based on trials rather than pupils. This is usually because the size of the trial-based percentage (very large or very small) indicated that detailed analysis on a pupil basis was not justified.

The following account is based on 273 pupils and 4,042 trials for 'Springload' and 274 pupils and 1,565 trials for 'Catalase'. Only the more significant results of this analysis will be commented on. The two tasks will be reported together, and the structure of this appendix is as follows:

A6.2 experimental design, and control of interfering variables

A6.3 operationalization of variables

A6.4 recording of data

A6.5 pupils' views of the design they had used

A6.6 variations in performance across examination entry groups

A6.2 Experimental design

The requirement in each of these two tasks is to determine the effect of the independent variable (spring length and width in 'Springload'; amount of liver sludge and amount of hydrogen peroxide in 'Catalase') on a third dependent variable. The basic requirement in the sequence of trials undertaken is to separate the two independent variables. The minimum requirement in each case is a pair of trials with one variable constant and the other altered, and the converse of this. Thus three trials would suffice. It might be argued that an effective approach to the independent variables requires a minimum of three separate values to have been used. It was considered that this is more appropriately regarded as an aspect of variable operationalization, and it is therefore discussed under that heading. The percentages of pupils setting up at least the minimal design requirements during the course of their experiments were:

'Springload' 92 per cent
'Catalase' 45 per cent

It is not difficult to suggest reasons for the striking difference here. Those who used all nine springs in 'Springload' (36 per cent of pupils) generated an experimental design automatically perfect in this respect. Moreover the variables width and length are operationalized (indeed their relationship is modelled) in the spring layout in 'Springload'. In 'Catalase' the variable relationship is less evident, and the situation is more open-ended.

Cross-tabulations of pupil performance on effective testing of width and length, and hydrogen peroxide and liver sludge are shown in Table A6.1 (p157).

Thus it appears that this aspect of variable handling/experimental design is achieved in an integrated rather than piecemeal fashion.

Figure A6.1 *'Catalase' question and apparatus*

When liver is added to hydrogen peroxide a gas is given off. The amount of gas given off may depend on several things. In this experiment you have to find the answer to this question:

> What affects the amount of gas given off?
> The amount of hydrogen peroxide used?
> Or the amount of liver sludge?

You can use any of the things in front of you. Choose whatever you need to answer the question.

Make a clear record of your results here so that someone else can understand what you have found out.

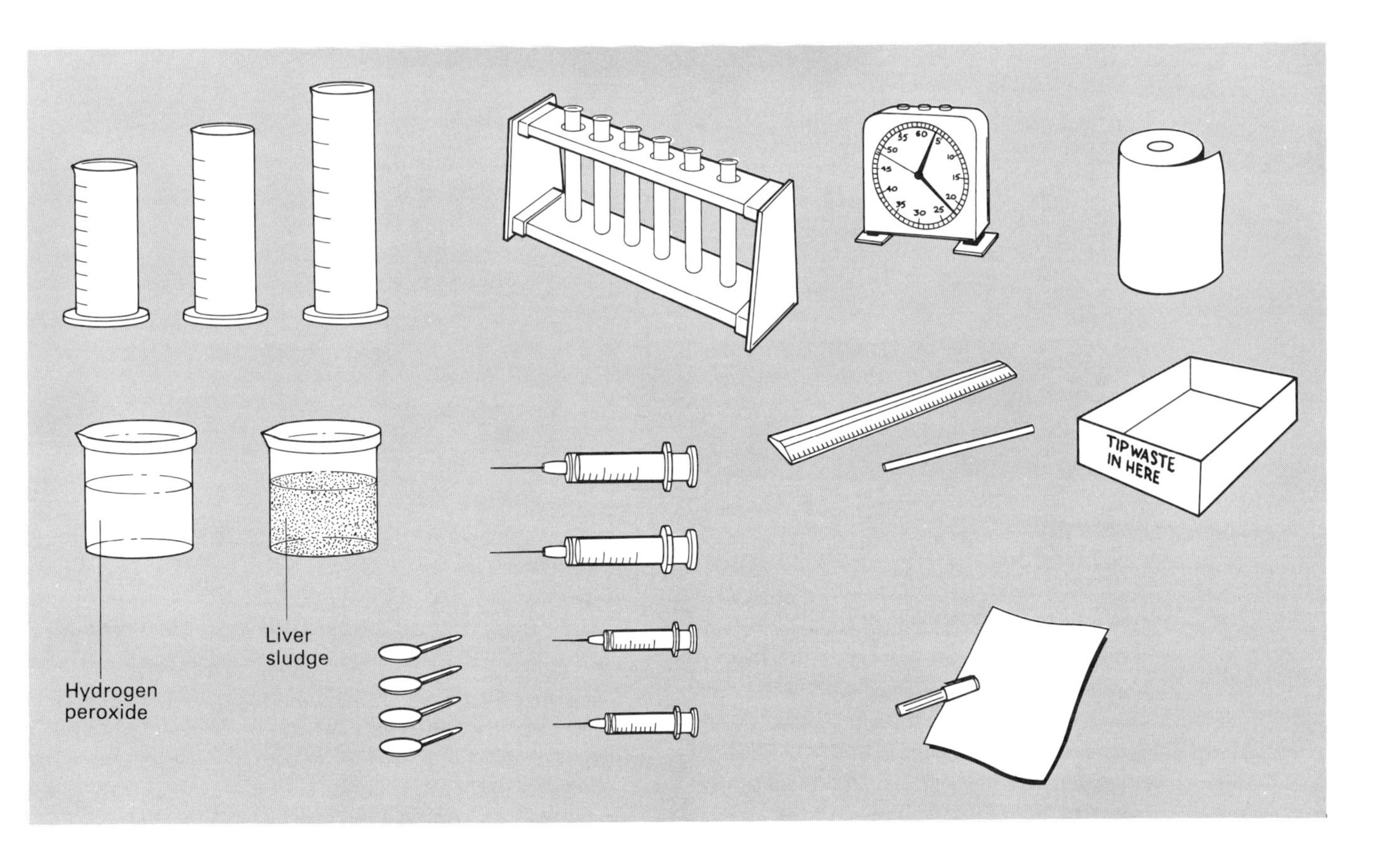

Figure A6.2 *'Springload' question and apparatus*

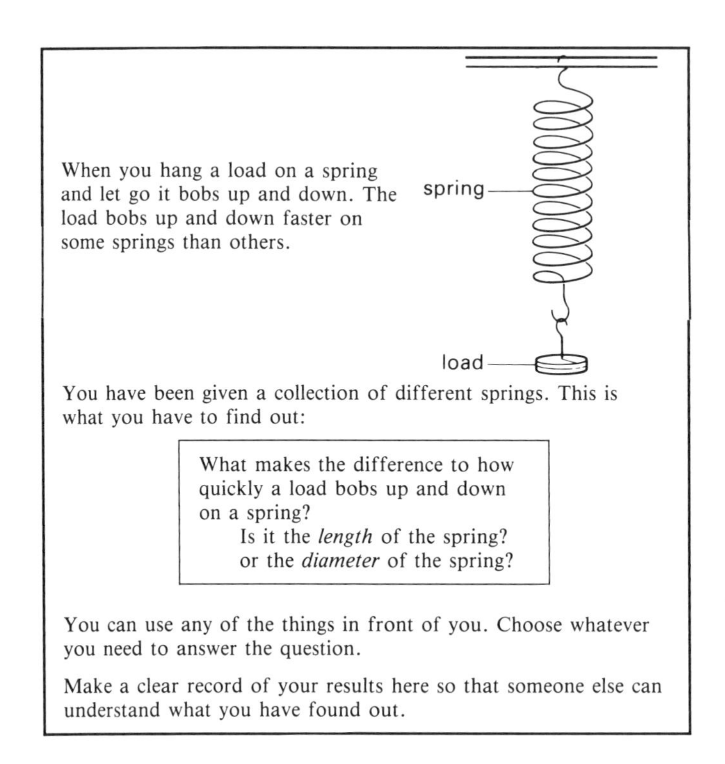

When you hang a load on a spring and let go it bobs up and down. The load bobs up and down faster on some springs than others.

You have been given a collection of different springs. This is what you have to find out:

> What makes the difference to how quickly a load bobs up and down on a spring?
> Is it the *length* of the spring?
> or the *diameter* of the spring?

You can use any of the things in front of you. Choose whatever you need to answer the question.

Make a clear record of your results here so that someone else can understand what you have found out.

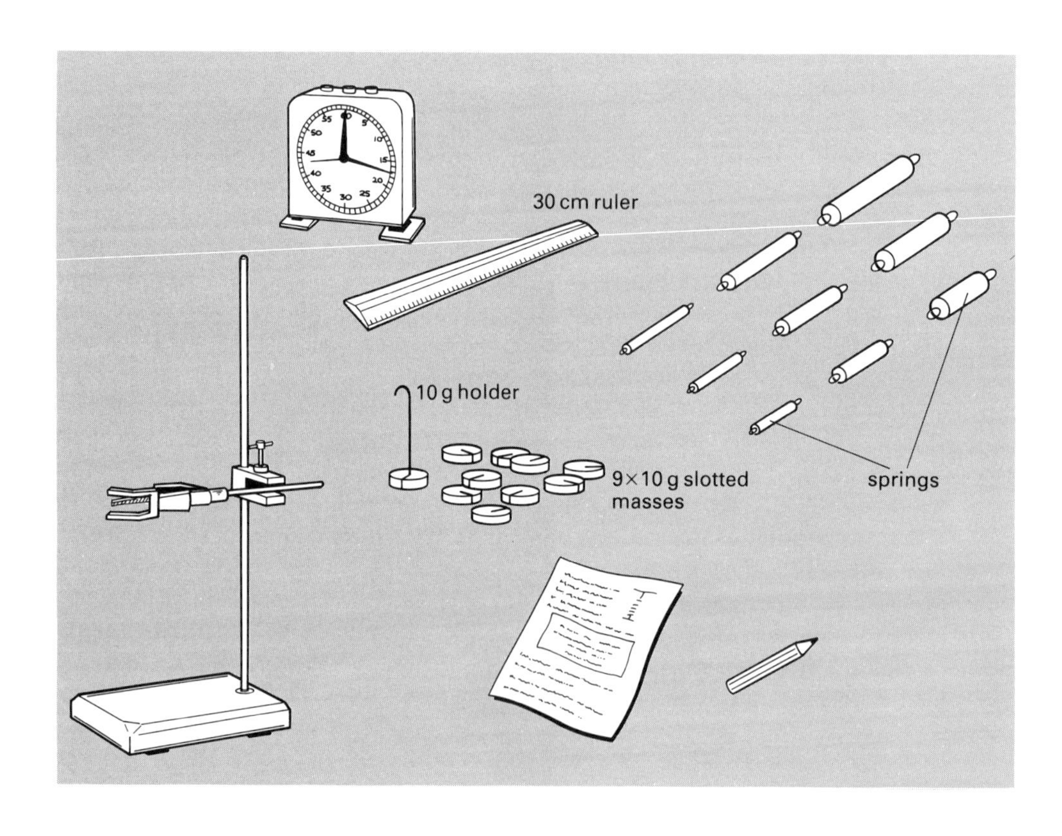

Table A6.1 *Performance on handling of independent variables*
(Percentages performing tests indicated)

SPRINGLOAD		Length tested					
		Yes			No		
		All	Boys	Girls	All	Boys	Girls
Width tested	Yes	92	89	95	4	5	3
	No	2	2	2	2	1	1
CATALASE		Hydrogen peroxide tested					
		Yes			No		
		All	Boys	Girls	All	Boys	Girls
Liver sludge tested	Yes	45	47	43	4	4	6
	No	7	3	9	44	46	42

Control of variables

Assessing the control of other variables in the two tasks presents problems. In 'Springload' the most obvious variables requiring control are the load used, the extent of initial displacement and the method of initial displacement. Seventy-eight per cent and 84 per cent of pupils respectively were judged by assessors to have used a consistent approach to the latter two. Approaches to the load used present problems because of the tendency of a significant minority of pupils (15 per cent) to treat the task in whole or part as related to Hooke's Law and for others to vary the load systematically. It is best approached by looking at the impact of requiring control of load on the experimental design discussed in the previous section. With this requirement, and based in general on the most frequently used load, the percentage achieving the minimal design for 'Springload' is 84 per cent (compared with 92 per cent above) suggesting a high level of control of this variable. Similarly, in 'Catalase' the variable most obviously requiring control is method of mixing. Supervisors assessed 45 per cent as having controlled mixing. Of the remainder, 45 per cent did not undertake any mixing, so that the question of its control may not have arisen. Control of a given variable appears to depend strongly on its perceived significance and the theoretical framework in which it is set. The control of variables which were considered as significant appears to have occurred at a high level.

A6.3 Operationalization of variables

The expected dependent variables were 'rate of bobbing' and 'height of "froth"' respectively. In 'Springload' considerable numbers perceived it differently. Fifteen per cent of pupils undertook a systematic measurement of spring extension, with seven per cent undertaking this *as well as* measuring rate of bobbing. The remaining eight per cent appear as having failed to measure rate of bobbing quantitatively in the table below, though we see that this stems from their perception of the task. This demonstrates the effect of previous experience of the physical situation on perception of the variables as presented.

The main interest in pupils' operationalizing of the dependent variable was the extent to which this had been done quantitatively rather than qualitatively. In addition the extent of a shift from one to the other during the course of the task was assessed. In 'Springload' quantitative approaches were taken to mean either counting the number of bounces in a fixed time, or the time for a fixed number of bounces. (Pupils adopting a quantitative approach split approximately in the ratio 2:1 between these approaches.) In 'Catalase' measurement of either the height or the volume of froth was acceptable. (Pupils adopting a quantitative approach split approximately 10:1 between these two approaches.) Clearly the complexity involved in a quantitative approach to 'Catalase' is much less than that for 'Springload'. However, so too is the difficulty of making a qualitative judgement of the independent variable. From a survey of the behavioural checklists the figures in Table A6.2 were obtained.

Table A6.2 *Performance on dependent variable operationalization*
(Percentage of pupils using approach indicated)

	'Springload'			'Catalase'		
	All	Boys	Girls	All	Boys	Girls
Quantitative throughout	12 }	13 }	12 }	26 }	29 }	22 }
Qualitative → quantitative	19 } 31	26 } 39	11 }	6 } 32	7 } 36	6 } 28
Quantitative → qualitative	2	2	2	3	3	4
Qualitative throughout	56	47	66	53	49	57
Other	11	12	9	12	12	12

It is striking that the percentage of quantitative and qualitative approaches is very similar for these very different variables. However, whereas a considerable learning effect (or at least improvement) is evident for 'Springload', this is not the case for 'Catalase'. We can perhaps see that the technical difficulties of variable operationalization are less significant than consciousness (or otherwise) of the need to operate quantitatively. Exposure to the physical situation appears to have had a noticeable effect in counteracting an initial failure to quantify the 'rate' variable in 'Springload'. We are of course unable to say what proportion of the remaining 70 per cent could under some circumstances operationalize the more complex variable.

Another noticeable aspect of these results is the sex difference in performance. In general, differences between boys and girls in the performance of these tasks are not significant. However, both 'Catalase' and 'Springload' display some evidence of differences. Table A6.2 can be dichotomized between effective quantitative approaches (the first two of those listed) and others. On this basis the difference in favour of boys in 'Catalase' is not quite significant at the 10 per cent level ($\chi^2 = 2.01$, df = 1). That in 'Springload' is significant at the one per cent level ($\chi^2 = 7.41$, df = 1). Moreover, the difference is almost entirely due to a shift from qualitative to quantitative approaches during the course of the investigation.

Other quantifiable aspects show a similar dependence on contingencies to that referred to above. Thus in 'Catalase', about 95 per cent of trials involved a quantitative approach to the volumes of hydrogen peroxide and liver sludge, whereas only 18 per cent of pupils undertook any systematic measurement of the dimensions of the springs in 'Springload' (ie measured all of the springs or all of those they used). This is probably due to the requirement to measure out quantities of materials in the former case.

Another aspect of the quality of operationalization in the contexts of both 'Catalase' and 'Springload' is the number of trials used to test the effect of a given independent variable. Here the situation in 'Springload' is more complex. It is not easy to distinguish design and operationalization at the lower performance levels. A pupil who fails to alter one variable in isolation from another may be said to have failed logically to distinguish it, and thus to function with it *qua* variable. Table A6.3 shows the results of a hierarchical analysis of behavioural checklists for 'Springload'.

Table A6.3 *Extent of variable operationalization in 'Springload'*

(Percentage of pupils using combinations of values indicated)

	All	Boys	Girls
All three widths and lengths used	58	57	59
At least two extreme widths and lengths used	17	16	18
At least two adjacent widths and lengths used	9	11	7
One or other variable not operationalized	16	16	17

These figures are included for comparison, though the unproblematic variable structure of 'Springload' must be borne in mind. In addition, the figures are based only on non-confounded tests, since otherwise there is no indication that values are consciously chosen.

A similar analysis of the summary checklists for 'Catalase' is shown in Table A6.4.

Table A6.4 *Extent of variable operationalization in 'Catalase'*

(Percentage of pupils using volumes indicated)

	All	Boys	Girls
At least 3 different volumes of each reagent used	53	53	52
At least 2 different volumes of each reagent used	35	40	33
One or other variable not operationalized	10	6	14

Because volumes of both reagents must be consciously chosen, these figures are *not* drawn from non-confounded tests only. This aspect of variable operationalization produces similar levels of performance for 'Catalase' and 'Springload'. The contrast with the poor figure for experimental design in 'Catalase' is clear. It can be seen that the great majority of pupils are in fact operationalizing both variables in an effective manner. However, in many cases they do not appear able to combine them into a coherent experimental design. About half of the pupils use the logical minimum of two values only for one or both variables. Furthermore, about 10 per cent of pupils in each case are not consistently altering the two characteristics of the situation intended as independent variables.

A6.4 Recording of data

A major interest for data recording in this category is concerned with its relationship to experimental design. The key difference is that between tabulated recording and other forms. The former provides an immediate model for the logic of the situation, the latter less clearly so, if at all. The relationship between experimental design and data representation is shown in Table A6.5.

Table A6.5 *The relationship between experimental design and data representation*

(Percentage of pupils using each combination of design and data representation)

		Tabulated	Ordered	Prose/random	No record
'Springload'					
Minimal design	Yes	19	40	28	5
	No	1	2	4	2
'Catalase'					
Minimal design	Yes	8	24	10	1
	No	2	18	28	3

It can be seen that tabulated data representation is only partially associated with adequate experimental design and that the former is not common. Tabulated and ordered data collection together are more closely associated with adequate design, but the latter does not usually reflect the variable structure of the task. The failure to reflect adequate experimentation in data recording may be due merely to lack of motivation, or to lack of writing/representational skills. However, it may also suggest that, despite success by behavioural criteria, pupils have not understood the underlying variable-based logic of the task which has been undertaken. The higher performance level in 'Springload' is probably a consequence of the physical modelling of the variable relationship in the springs.

A6.5 Pupils' views of the design they had used

In order to take the point made in the previous paragraph further, 'Springload' pupils were asked, after completion of the task, to state the springs they had used to test length and width. Their answers were analysed on the same basis as their actual use of springs. Ninety-four (34 per cent) pupils answered that they had used all springs, or some uninterpretable combination, for testing either length or width. Of the remainder, non-confounded testing of length and width was suggested as shown in Table A6.6. The percentages shown in Table A6.6 are for all pupils. The proportion of pupils who, indicating that they had used all of the springs, meant that they had carried out three separate non-confounded tests cannot be judged.

Table A6.6 *Response to questioning on method of testing length and width*
(Percentages performing tests indicated)

		Length tested	
		Yes	No
Width tested	Yes	33	13
	No	13	41

These figures indicate that the *conscious* testing of springs may have occurred at a much lower level than is suggested from the behavioural data. The figure is comparable with that observed in 'Catalase'.

A noticeable aspect of pupils' response was the tendency to 'test' for length by looking only at the longest springs. Pupils' responses were analysed on this basis, and it was found that, of those who failed to suggest a non-confounded test for length (and whose response to the question could be analysed) 56 per cent suggested that some combination of three longest springs be used. This may of course be a transposition of the logic of testing for width and length. Alternatively, it may indicate that the logic of testing by variable *manipulation* has not been achieved. Within this view, one would 'test' for length by looking at the properties of long (or perhaps short) objects. The effect was noticeably less marked for width, with only 31 per cent of those who failed to suggest a non-confounded test suggesting the use of the three widest springs.

Similar questions were put to pupils who had attempted 'Catalase'. In order to reduce question length it was necessary to invite a positive response. Thus pupils who did not explicitly state that amount of liver sludge needed to be held constant when testing for the effect of hydrogen peroxide (and vice versa) may have taken this as obvious. Responses are shown in Table A6.7.

Table A6.7 *Response to questioning on methods of testing hydrogen peroxide and liver sludge*
(Percentage giving indicated responses)

	Hydrogen peroxide varied and liver sludge controlled	Hydrogen peroxide varied only	Other
Liver sludge varied and hydrogen peroxide controlled	14	0	3
Liver sludge varied only	1	24	4
Other	5	4	46

It is clear that the responses to these separate questions are strongly related, a situation which parallels that for pupils' practical performance. The large percentage of pupils not expressing any sensible test in either case included many who in effect combined the two variables. Thus they suggested that the effect of liver sludge was to be observed by using a lot of it with a little hydrogen peroxide. Clearly this has parallels with, but is not identical to, the situation in 'Springload'. It was possible, for 'Catalase', to compare reported with actual performance on the design of the experiment. The results are shown in Table A6.8.

Table A6.8 *Performance in practical and oral responses to 'Catalase'*
(Percentages giving indicated responses or practical performance)

		Practical		
		Non-confounded test of one or both variables	Variation of both variables	Other
	Non-confounded test of one or both variables	18	3	1
Oral	Variation of both variables	13	11	1
	Other	24	23	7

These data show that performance on this aspect of the task matches oral accounts in about 33 per cent of cases. In the remainder of cases practical performance is almost always better. Two interpretations of the values below the diagonal are possible. Pupils may lack the language skills to give an account of their practical activity. Alternatively it may be that their own perception of this aspect of their activity differs significantly from that obtained using external 'behavioural' criteria. As has been suggested elsewhere in this appendix, there is some evidence of a mismatch between the pupils' perception of the structure of tasks and that implied within the variable-based analysis within which data are collected and performance reported.

A6.6 Variations in performance across examination entry groups

Within APU, survey data on pupils' likely examination entry are collected as a surrogate measure of ability. The categories used are:

1 8+ O-levels
2 6–7 O-levels
3 3–5 O-levels
4 1–2 O-levels
5 4+ CSE
6 0–3 CSE

In order to provide sufficient sample sizes the groups are amalgamated:

A 6+ O-levels (more able)
B 1–5 O-levels (average)
C CSEs only (less able)

Certain aspects of pupils' practical performance have been analysed according to these 'ability' groups, and the resulting data are presented in this section.

Pupil performance on experimental design was analysed into three categories: those undertaking a non-confounded test of both independent variables, those undertaking only one such test, and those undertaking no non-confounded tests. The results are shown in Table A6.9.

Table A6.9 *Handling of independent variables by pupils in different examination entry groups*
(Percentage of each group undertaking indicated test*)

Examination entry group	Non-confounded test of both variables		Non-confounded test of one variable		All testing confounded	
more able	98	**65**	1	**8**	1	**25**
average	91	**42**	6	**15**	3	**43**
less able	72	**28**	19	**10**	9	**62**

* Normal print refers to 'Springload', **bold** print to 'Catalase'.

The most obvious point here is the *contrast* in the trend between 'Catalase' and 'Springload'. Performance on 'Catalase', where setting up an appropriate experimental design requires considerable care and abstraction, falls away more rapidly, particularly if the first two approaches are amalgamated. In addition, the more able exhibit a greater stability of response across the two tasks. It is noticeable, however, that in the more demanding situation in 'Catalase' 28 per cent of those likely to be entered for CSEs only produced a full non-confounded test, while 25 per cent of those entered for 6+ O-levels produced a fully confounded test. This overlap is greater than might be expected.

These data are of course based on pupils' practical activity. It was seen in the previous section that their verbal account of this aspect of the tasks they had undertaken diverged considerably from their actual activity. The analysis of these verbal accounts by examination group is shown in Table A6.10.

Table A6.10 *Pupils' verbal accounts of the design they had used*
(Percentage of each examination entry giving indicated response*)

Performance group	Both independent variables tested		One independent variable tested		Neither independent variable tested	
more able	45	**28**	15	**11**	41	**60**
average	35	**12**	27	**6**	38	**82**
less able	19	**6**	36	**8**	46	**86**

* Normal print 'Springload'; **bold** print 'Catalase'.

There is some indication here that using a verbal account has a differential impact on the performance groups. The least able do less well than the most able (in relation to their respective practical performances) in both tasks. Thus, in 'Catalase' the decline in those conducting a fully non-confounded test in practice is from 65 per cent to 28 per cent, whereas it is from 28 per cent to 6 per cent in verbal accounts. Moreover, while the explanation of the discrepancy between practical and verbal performance for the more able in 'Catalase' may involve their tendency to assume the control of the second independent variable, this explanation is not available for the least able. The majority of this group indicated more complex responses (often explicitly confounded).

Controlling contingent variables is considered to be a component of experimental design. The variables requiring such control were considered to be method and position of release for 'Springload', and amount of mixing for 'Catalase'. Performance on these aspects is shown in Table A6.11 (p161).

It is noticeable that this represents a relatively high level of performance for the less able. Both aspects of the task in 'Springload' show a tendency for a more limited or less systematic approach on the part of lower ability groups, rather than total failure to take account of the variable. The situation in 'Catalase' is complicated by the fact that the need to mix the two liquids can be reduced by the mixing which occurs when one is added to the other, which in turn depends on the relative amount being added. More sophisticated understanding of this process (and of the overall situation) may have led to the large proportion of more able pupils who selectively controlled the mixing of the reactants.

Operationalizing the dependent variable

Peformance on this was divided into groups of those exhibiting wholly quantitative approaches, movement from a qualitative to a quantitative and other approaches. The results are shown in Table A6.12.

Table A6.11 *Control of variables by examination entry group*
(Percentage of each group controlling indicated variable)

	'Springload'				**'Catalase'**	
	Method of release controlled		Position of release controlled		Amount of mixing controlled	
	throughout	sometimes*	throughout	sometimes	throughout	sometimes
more able	95	4	94	1	**43**	**15**
average	90	1	85	2	**42**	**5**
less able	78	16	79	13	**44**	**2**

* 'Control sometimes' indicates a systematic attempt to control the variables stated within a set of runs which *did not* constitute a full experiment.

Table A6.12 *Operationalization by examination entry group*
(Percentage of each group operationalizing as shown*)

	Quantitative		Qualitative → quantitative		Other	
more able	18	**40**	28	**11**	54	**48**
average	16	**23**	17	**2**	67	**75**
less able	5	**18**	11	**5**	84	**77**

* Normal print 'Springload'; **bold** print 'Catalase'.

The relative ease of quantitative measurement in 'Catalase' evidently had a limited effect on the performance of the least able. It appears that the more able were more inclined to take a quantitative measurement in the first instance *and* to move from a qualitative to a quantitative measurement. It is noticeable that, in the more complex situation of 'Springload', it was the willingness/ability to *develop* a quantitative approach which differentiated the most able. However, the movement to such an approach was made by greater numbers of the less able in 'Springload' (where the need was clear) than in 'Catalase' where it was relatively easy to avoid.

Style of data recording

Results of the analysis into tabulated, ordered, prose and random are shown in Table A6.12. Again the polarization between the most able and the remainder is very strong. Even for the former, tabulated data recording is generally rare, and the major shift is between ordered data recording and prose. The tendency to present data separately (ie not as a component of a prose account) is very great for the most able and shows a noticeable decline for the least able, but superimposed on this are the greater problems apparently presented by the more abstracted situation of 'Catalase'.

Table A6.13 *Style of data record by examination entry group*
(Percentages of each group using indicated record*)

	Tabulated		Ordered		Prose		Random		No Record	
more able	35	**22**	50	**51**	10	**18**	2	**4**	2	**4**
average	19	**7**	45	**41**	19	**46**	10	**5**	7	**1**
less able	6	**2**	36	**37**	46	**47**	3	**1**	9	**12**

* Normal print 'Springload'; **bold** print 'Catalase'.

Appendix

7

List of science concepts and knowledge at age 15

A Interaction of living things with their environment

A1 Interdependence of living things

Virtually all organisms are dependent on the presence and activities of other organisms for their survival.

Green plants use energy from the sun to make food by 'photosynthesis'. During this process plants produce oxygen. Plants and animals use oxygen in respiration.

Some animals eat plants and some eat other animals, but all animals ultimately depend on green plants for food. This relationship can be illustrated by a 'food web'.

Any alteration in one part of the 'food web' may affect many other parts of the 'food web'.

An alteration in the 'food web' can be as a result of a change of balance between 'consumers' and 'producers', or of a change in the inorganic environment.

Organisms live in 'communities' in which each organism has a place in which it is best adapted to survive.

Competition and predation tend to maintain the balance of populations within a 'community'.

A2 The physical and chemical environment

The physical and chemical conditions necessary to life occur on or near the earth's surface.

The composition of air tends to remain relatively stable as a result of cyclic processes involving living things.

Water is essential both as a raw material and as the 'solvent' involved in transport and chemical processes within living things.

Soil is a mixture that includes rock particles, humus, water, air and living things.

The return of substances to the soil by death and decay of living things helps to maintain soil fertility.

Changes in the physical environment due to seasonal cycles are often matched by changes or events in the living world, such as fruiting or mating.

A3 Classification of living things

There are many different plants and animals showing a variety of ways of carrying out life processes.

There are many different criteria that can be used to sort living things, eg structure, behaviour, etc.

Plants are distinguished from animals by a number of criteria including cell structure and method of obtaining food.

Plants are classified into two major groups–flowering and non-flowering.

Animals are classified into two major groups–invertebrates and vertebrates.

There are five main groups of vertebrates: fish, amphibia, reptiles, birds and mammals.

A theory about how organisms have 'evolved' is currently used to classify the living kingdom.

A4 Physical and chemical principles needed to interpret phenomena

Diffusion in living organisms occurs in an aqueous medium. Water particles move by diffusion as do particles of substances dissolved in water.

Osmosis is a special case of diffusion taking place across a 'selectively permeable' membrane which usually allows the passage of water particles only.

The area of its surface affects the gain or loss of energy and matter from an organism or cell.

The ratio of surface area to the mass of an organism is critical for its survival and will reflect its relationship with its environment.

Animals which have a stable body temperature have a number of mechanisms for the control of loss of heat energy.

B Living things and their life processes

B1 The cell

The cell is the basic unit of most living things.

A 'typical' cell has a nucleus, cytoplasm and a cell membrane.

The nucleus is the controlling centre of the cell and contains 'hereditary' material.

The cell membrane acts as a 'selectively permeable' barrier.

Plant cells have a cell wall outside the cell membrane.

Cells show a wide range of adaptation to the particular function they perform.

An organism may be formed from one or many cells.

In multicellular organisms, cells are usually grouped to form tissues and tissues grouped to form organs or systems.

Normal cell division, 'mitosis', results in two identical cells and leads to either growth or asexual reproduction.

B2 Nutrition

Food is used by all organisms as a source of energy and of raw materials for growth and reproduction.

Green plants take in simple substances from their environment and build them into more complex substances.

All food substances must be dissolved before the cells can use them.

Natural foods contain fats, proteins and carbohydrates.

In humans, an adequate diet will contain a sufficient and balanced amount of these foodstuffs as well as water, vitamins, minerals salts and roughage.

Digestion is the process in which foods taken into the body are broken down by mechanical and chemical means into simple soluble substances for absorption and transport.

Surplus food is stored in an insoluble form as fat or carbohydrate.

Enzymes change the rate of certain chemical reactions under specific conditions.

All organisms are 'structurally adapted' to the kind of food they need.

The structure of plants reflects their ability to manufacture organic food using absorbed light energy.

All living things produce waste materials in carrying out life processes.

B3 Respiration

Most living things take in oxygen from their surroundings to be used in the process of respiration.

All living things respire in order to make use of energy stores in food.

In respiration, energy is transferred from organic food material with the release of carbon dioxide.

In the case of 'aerobic' respiration, water is also produced.

Many living things have special organs which enable gas exchange to take place effectively.

Gas exchange organs have a large surface area for their volume, are damp, and often employ 'mass flow' aids.

B4 Reproduction

Living things produce offspring of the same species as themselves.

Asexual reproduction involves one parent and the offspring are identical to that parent.

There are many different types of asexual reproduction but they all involve 'mitotic' cell division.

Sexual reproduction involves two parents and generally results in offspring that differ from both parents and from each other. This leads to variation.

Sexual reproduction involves the production of 'gametes', ova and sperm or pollen, produced by a special type of cell division ('meiosis'), and their fusion to form a zygote.

The human sexual cycles and changes are controlled by hormones.

The life cycle of all organisms repeats itself every generation.

B5 Sensitivity and movement

All living organisms have means of receiving information from their environment.

Special organs/cells are concerned with receiving different kinds of 'stimuli'. These are called sense organs/cells.

In humans, the senses include sight, touch, hearing, taste and smell.

The senses developed by an organism show an adaptation to its particular way of life.

The response to a 'stimulus' often results in movement or, in higher plants, in directional growth.

Voluntary movement in many animals is brought about by contraction of muscles attached to a skeleton.

C Force and field

C1 Movement and deformation

The average speed of an object is found by dividing the distance moved by the time taken.

When an object travelling in a straight line changes its speed it is said to accelerate. The size of the acceleration is the change in speed divided by the time.

An object which is still or moving with uniform speed in a straight line has no unbalanced forces acting on it.

An unbalanced force makes an object accelerate and the acceleration is greater when the force increases. For a given unbalanced force, the acceleration is smaller for larger masses.

Rotation is produced when an object has a pair of parallel forces which are not in line acting on it; the turning effect, or 'moment' of a force about a point is the magnitude of the force multiplied by the distance of the line of action of the force from the point.

Objects are deformed by and can break under the action of 'opposed' forces. The force which an object can stand before breaking depends on its shape as well as the material it is made of.

If materials recover their original shape after a force which caused deformation has been removed, the materials are called elastic.

C2 Properties of matter

Different substances have different masses for equal volume; the mass of a unit volume of a substance is its density.

If the volume of a given mass increasees, its density decreases.

Pressure is the magnitude of the force per unit area acting on a surface.

Pressure at a point in a gas or liquid depends on the depth of the point below the surface and the density of the gas or liquid.

Pressure at a point within a gas or liquid acts equally in all directions if there is no movement in the gas or liquid.

Particles of a substance are in constant motion. Diffusion of a substance is due to the random motion of individual particles (which may be molecules, atoms or ions).

If the pressure of a fixed mass of gas is increased, its volume decreases provided the temperature stays the same.

The pressure exerted by a gas on a surface is the result of continual bombardment of the surface by many particles.

Most substances expand as their temperature increases.

Rigid objects completely immersed in a gas or liquid displace a volume of gas or liquid equal to their own volume.

There is an upward force on an immersed object which is equal to the weight of gas or liquid displaced; a floating object displaces a weight of gas or liquid equal to its own weight.

Surface tension and capillarity may be accounted for in terms of intermolecular forces.

C3 Forces at a distance

Magnets attract and repel other magnets and attract magnetic substances.

The region in which a magnetic effect can be detected is called a magnetic field.

There is a magnetic field surrounding the Earth.

Magnetism is induced in some materials when they are placed in a magnetic field.

An electric current in a conductor produces a magnetic field round it.

There is a force on a current-carrying conductor in a magnetic field so long as the conductor is not parallel to the field.

The force on a current-carrying conductor in a magnetic field increases with the strength of the field and with the current.

Relative motion of a conductor and a magnetic field can be used to produce an electric current.

Electric charges are separated when certain materials are rubbed against one another.

Similarly charged objects repel each other, and oppositely charged objects attract each other. The force between such objects is increased by bringing the objects closer and by increasing the charges on them.

C4 The Earth in space

The weight of an object on the Earth is the force with which it is attracted to the Earth.

The weight of an object may vary from place to place, but its mass does not change.

The weight of an object acts through its centre of gravity. The position of the centre of gravity affects the stability of the object.

The apparent movements of the sun, moon, planets and stars follow a regular pattern.

The Earth spins on its axis; this gives rise to night and day.

The Earth revolves in an orbit around the sun with its axis tilted with respect to this orbit; this accounts for seasonal changes.

The moon orbits the Earth as a natural satellite.

D Transfer of energy

D1 Work and energy

There is a variety of sources of energy such as fuels (including food), other chemicals, deformed springs, capacitors and objects at a height.

Energy can be changed from one form to another but can never be created or destroyed. At each change some energy becomes less available for doing useful work.

Moving objects have energy which is transferred when they are stopped.

An object increases its 'potential energy' when it is raised from its original position to a greater height above the Earth's surface.

Work is done (energy is transferred) when a force moves its point of application; work is measured by the product of the force and the distance moved in the direction of the force.

The hotter the substance is, the more energy its particles have.

Different substances conduct heat at different rates. Conduction of heat can be explained in terms of the kinetic theory of matter.

The expansion of material on heating can be explained in terms of the additional energy of motion of its particles.

D2 Current electricity

A complete circuit of conducting material is necessary for a steady current to flow between the terminals of a battery or a d.c. power supply.

Electric current is a flow of charged particles.

Some materials conduct electricity better than others. Bad conductors are known as insulators.

Electric currents add up 'algebraically' at a junction in a circuit.

D3 Waves

Objects are seen because of the light which they give out or reflect.

The path of light travelling in a uniform medium can be represented by straight lines.

Light is reflected regularly at a mirror-like surface.

Light is 'refracted' at the boundary between two different media which both transmit light.

'Refraction' and reflection give rise to the formation of images.

Sound can be heard when objects vibrate in a medium.

Sound requires a medium for its transmission.

In general, the loudness of a sound is increased when the amplitude of vibration of the source is increased.

When the frequency of vibration is increased, the note sounds higher.

Different colours of visible light have different wavelengths. Other types of radiation share some of the properties of light and can be located on an extended scale of wavelengths.

E The classification and structure of matter

E1 States of matter

In general a substance can be classified either as a solid or a liquid or a gas.

The behaviour of substances can be explained if it is assumed that matter is made of minute particles.

Solids have a definite shape and volume. They behave as if their particles are held together in regular arrangements.

Liquids have a definite volume and surface but no fixed shape. They behave as if their particles are closely packed but free to move.

Gases have no definite volume or shape. They behave as if their particles are free to move independently of each other.

A change in state does not involve a change in the chemical composition of the substance.

Changes of state caused by heating can be reversed by cooling and vice-versa.

Changes of state always involve a transfer of energy.

E2 Pure substances

A pure substance may be obtained from a mixture using one of several techniques, including evaporation, distillation, chromatography, filtration, sublimation and crystallisation.

A pure substance is recognised by its characteristic chemical and physical properties (at STP) eg mp bp, density and behaviour with other substances.

Different pure substances need different amounts of heat to give the same rise in temperature in a given mass.

Pure substances may be classified into elements and compounds.

An element is made up of only one kind of atom, ie of fixed atomic number.

A compound is formed when two or more different elements are chemically combined.

E3 Metals and non metals

Materials can be classified into groups in many different ways. One way is by sorting into metals and non-metals.

In general, metals can be distinguished from non-metals by their characteristic physical properties. These include high mp and bp, shiny appearance and conduction of heat and electricity.

Another way of distinguishing metals from non-metals is by their characteristic chemical properties. These may include the nature of the oxide and the formation of positive ions.

Some metals do not exhibit all these characteristic properties.

Some elements show both non-metallic and metallic characteristics.

E4 Acids and bases

Compounds can be classified into groups by their different properties. Acids and bases are two such groups.

Acids and bases have a characteristic effect on the colour of indicators.

Bases which are soluble in water are called alkalis.

In aqueous solution the degree of acidity depends on both the substance and its concentration.

The degree of acidity is expressed on a pH scale from 0 to 14.

A neutral solution has a pH value of 7; acidic solutions have pH values less than 7, and alkaline solutions have pH values greater than 7.

A neutral solution can be obtained by adding an acid to an alkali until the pH value is 7.

E5 Periodic table

Families of elements with similar chemical behaviour can be identified.

Patterns and trends exist in the behaviour of the elements.

The characteristic properties of the elements show a periodic dependence on atomic number.

E6 Atomic model

The atom is the smallest, characteristic uncharged particle of an element.

The behaviour of atoms can be explained if it is assumed that they are composed of a heavy positive nucleus surrounded by negative electrons.

Atoms are characterised by a symbol and the atomic number (number of protons).

An uncharged bonded group of two or more atoms is called a molecule.

An ion is an atom or bonded group of atoms bearing electric charge(s), positive or negative.

F Chemical interactions

F1 Solutions

Some substances dissolve in water; others do not, but may dissolve in other liquids.

A liquid which will dissolve a substance to form a solution is called a solvent.

The substance which dissolves in the liquid is called a solute.

At a given temperature and pressure, the mass of solute which will dissolve in a given volume of solvent is limited and fixed.

For most solids, solubility increases with increasing temperature.

A saturated solution of a solute in a solvent is one in which no more of that solute will dissolve, at a given temperature and pressure.

The mass of solute which is dissolved in a given volume of solvent determines the concentration of the solution.

F2 Reactivity

A more reactive element can be used to extract a less reactive element from one of its ores or compounds.

The reactions of metals with water or dilute acid can be used to place them in a reactivity series.

The reaction of one metal to replace another from an aqueous solution of its salts may be predicted from this reactivity series.

The polarity of electrodes in a chemical cell can be deduced from the position in this reactivity series of the elements from which they are made.

F3 Properties of a chemical reaction

A chemical reaction occurs when one or more different substances are formed from one or more original substances.

Most chemical reactions are initiated by an input of energy.

The total mass of reacting substances in any chemical reaction is the same as the total mass of products.

Chemical reactions take place at varying and various rates.

The rate of a chemical reaction may depend on several factors, eg concentration, pressure and temperature.

A catalyst is a substance which changes the rate of a chemical reaction without being used up in the reaction.

F4 Some chemical reactions

On heating some compounds change colour due to loss of water. Often these changes are easily reversed.

When elements react with oxygen only, they usually form oxides; this is an example of an oxidation reaction.

When a compound changes by losing oxygen, this is an example of a reduction reaction.

The oxidation of a metal by atmospheric oxygen is an example of corrosion.

Fuels such as coal and oil are formed by the gradual decay of plant remains under high pressure.

Large amounts of energy can be transferred from these fuels when they react with oxygen. This is an example of a combustion reaction.

When an acid reacts with an alkali or a base, a salt and water are formed.

The process of electrolysis involves ('electron transfer') reactions which are used extensively in industry.

Appendix

8

Practical test supervisors, 1984 monitoring

Group practical tests

Mr J. Anthony	Lewis Boys Comprehensive School, Pengam, Gwent
Mr J. Atkins	Darton High School, Kexborough, S Yorkshire
Mr R. L. Austin	Garth High School, Morden, Surrey
Mr B. Baker	(Retired) Saltash, Cornwall
Mr J. Bassett	Comber High Shcool, Newtownards, N Ireland
Dr R. N. Bradley	Dukinfield High School, Stockport, Cheshire
Mr R. W. Broyd	Minchenden School, Southgate, London N14
Mr T. Bryson	Strabane High School, Strabane, Co Tyrone, N Ireland
Mr D. C. A. Bull	de Stafford County Secondary School, Caterham, Surrey
Mrs A. S. Curry	The Misbourne School, Great Missenden, Buckinghamshire
Mrs M. M. Darkins	Nantglo Senior Comprehensive School, Nantglo, Gwent
Mr J. M. Dear	Drayton Manor High School, Hanwell, London W7
Mr D. Dimmock	Culey Green School, Sheldon, Birmingham, W Midlands
Mr B. Dingle	Tapton School, Sheffield, S Yorkshire
Dr T. Evans	Barry Boys Comprehensive School, Barry, S Glamorgan
Miss S. Farnfield	Hammersmith School, White City, London W12
Mr K. B. H. Herbert	The Beaufort School, Tuffley, Gloucestershire
Miss V. P. Hewetson	Joseph Rowntree School, York, N Yorkshire
Mrs A. Huard	Billericay School, Billericay, Essex
Dr M. Isles	Cantonian High School, Cardiff, S Glamorgan
Mr G. Jones	David Hughes School, Menai Bridge, Gwynedd
Mr J. Lane	Wellsway School, Keynsham, Bristol, Avon
Mr F. McCann	St Colmcille's High School, Crossgar, Downpatrick, N Ireland
Miss R. McCann	St Louis Grammar School, Ballymena, Co Antrim, N Ireland
Mr B. McCoubrey	Ballymoney High School, Ballymoney, Co Antrim, N Ireland
Mr R. McKearney	St Ciaran's High School, Dungannon, Co Tyrone, N Ireland
Dr L. McLoughlin	St Michael's Senior High School, Lurgan, Craigavon, N Ireland
Mr R. Meredith	The Woodhouse School, Tamworth, Staffordshire
Mr I. Morgan	Ysgol Gyfun Maes-Yr-Yrfa, Llanelli, Dyfed
Mr W. P. Morris	Littleport Village College, Littleport, Ely, Cambridgeshire
Ms D. O'Harte	Sacred Heart Grammar School, Newry, Co Down. N Ireland
Mr I. Richardson	Fallibroome High School, Upton, Macclesfield, Cheshire
Mr G. Roberts	Darland School, Rossett, Clwyd
Mr K. A. Ross	Norton Secondary School, Stockton, Cleveland
Dr C. J. R. Sneyd	The West Bridgford School, West Bridgford, Nottinghamshire
Mr S. Walker	Olchfa School, Sketty, Swansea, W Glamorgan
Mr D. J. Westcott	Queen Elizabeth's School, Crediton, Devon
Mr R. J. Wilcox	Smestow Comprehensive School, Wolverhampton, W Midlands

Individual practical tests

Mrs A. C. Archer	St Michael's School, Watford, Hertfordshire
Mr B. Brand	Eastbourne Comprehensive School, Darlington, Co Durham
Miss W. Brown	Loreto College, St Albans, Hertfordshire
Mr M. S. Cleverly	St Ilan Comprehensive School, Caerphilly, Mid-Glamorgan

Mrs C. N. Cobb	Heolddu Comprehensive School, Bargoed, Mid-Glamorgan
Mrs J. Coley	Huxlow School, Wellingborough, Northamptonshire
Mr M. R. Croft	Edge End High School, Nelson, Lancashire
Mr R. E. Dawson	Madeley Court School, Madeley, Telford, Shropshire
Mr J. W. Eaden	Kaskenmoor School, Oldham, Lancashire
Mr J. F. Faulkes	Fairfax Community School, Bradford, W Yorkshire
Mrs D. A. Gaskill	Welshpool High School, Welshpool, Powys
Mr G. W. Gaskill	Llanfyllin High School, Llanfyllin, Powys
Mrs G. Gotto	(Retired) Belfast, N Ireland
Mrs I. Hall	Cross Hall High School, Ormskirk, Lancashire
Mr S. Harrison	John Lea School, Wellingborough, Northamptonshire
Mrs A. Haycock	Cookstown High School, Cookstown, Co Tyrone, N Ireland
Mr W. F. Hinwood	Newport Lliswerry Comprehensive School, Newport, Gwent
Mr R. Jackson	Burford School, Burford, Oxford, Oxfordshire
Mr I. Johnson	Longfield Comprehensive School, Darlington, Co Durham
Mr R. F. Johnson	Gordano School, Portishead, Bristol, Avon
Mr C. Jones	Ercall Wood School, Wellington, Telford, Shropshire
Mr G. G. Jones	Scawsby Ridgewood Comprehensive School, Scawsby, S Yorkshire
Mr J. E. Jones	The Radclyffe School, Chadderton, Oldham, Lancashire
Mr P. Laithwaite	Oxford Teachers' Centre, Cowley, Oxfordshire
Mr R. Lane	Oakbank Grammar School, Keighley, W Yorkshire
Mr E. Mallon	St Patrick's High School, Dungiven, Londonderry, N Ireland
Mr H. McCullough	Dunluce High School, Bushmills, Co Antrim, N Ireland
Mr P. McGuckin	St Joseph's High School, Coalisland, Co Tyrone, N Ireland
Mr D. McKnight	Highdown School, Reading, Berkshire
Mr M. G. McQueen	Kelsey Park School for Boys, Beckenham, Kent
Mr W. Miles	Llantarnam Comprehensive School, Cwmbran, Gwent
Mr K. Parker	Cymer Afan Comprehensive School, Port Talbot, W Glamorgan
Dr E. K. Porter	Turnpike School, Newbury, Berkshire
Mr G. P. Richards	St Josephs RC Comprehensive School, Port Talbot, W Glamorgan
Mr D. Robbins	Westwood St Thomas School, Salisbury, Wiltshire
Mrs W. Swarbrick	Steyning Grammar School, Steyning, W Sussex
Mr A. G. Thomas	The Littlehampton School, Littlehampton, W Sussex
Mr B. Thompson	St Patrick's College, Knock, Belfast, N Ireland
Mr M. Todd	Garvagh High School, Garvagh, Coleraine, N Ireland
Mr D. A. Walsh	Edlington Comprehensive School, Doncaster, S Yorkshire
Dr A. Waters	Hayes School, Hayes, Bromley, Kent
Mr P. Winters	Convent Grammar School, Strabane, Co Tyrone, N Ireland

Appendix

9

Membership of Groups and Committees

A9.1 The 1984 Monitoring Teams

University of Leeds

Director	Fred Archenhold
Technical Director	Roger Hartley
Research and Development (age 15)	Richard Gott (Deputy Director) Angela Davey Reed Gamble Geoff Welford
Data Analysis	Sandra Johnson (Deputy Director) John Bell Nasrin Khaligh Iain Watson
Secretarial staff	Glynis Wilkinson Nadine Hannam Elizabeth Lodge

King's College, London

Director	Paul Black
Research and Development (ages 11 and 13)	Wynne Harlen (Deputy Director) Patricia Murphy Tony Orgee David Palacio Terry Russell Beta Schofield
Secretary	Peggy Walker

A9.2 APU Steering Group on Science (January 1987)

Mr A. G. Clegg HMI (Chair)	Professional Head of the APU
Mr W. F. Archenhold	Director, Science Monitoring Team, University of Leeds
Professor P. J. Black	Director, Science Monitoring Team, King's College London
Mrs S. Dean	St Martin's College, Lancaster
Dr N. B. Evans	HM Inspectorate (Wales)
Mr A. Giles	British School Technology, Trent Polytechnic
Mr E. O. James	Deputy Head, Southlands School
Professor R. F. Kempa	Department of Education, University of Keele
Dr W. J. Kirkham	Director, Secondary Science Curriculum Review
Mr E. R. B. Little	HM Inspectorate
Miss R. Jarman	Department of Education for Northern Ireland
Mr H. D. Wilcock	Headteacher, Paganel Junior School, Birmingham

Past members

Mr B. Barker 1978–1981 St Audrey's School, Hatfield
Mr T. A. Burdett HMI 1980–1982 Chairman
Professor D. Child 1982–1983 Director, Science Monitoring Team, Leeds
Mrs H. Davies 1984–1985 Bradford LEA
Mr J. Graham HMI 1979–1980 Chairman
Mr A. R. Hall 1978–1982 Sir Joseph Williamson's School, Rochester
Sister M. Hurst 1978–1982 Schools Council
Mr J. R. Jeffrey 1981–1984 Pocklington School
Professor P. Kelly 1978–1981 Co-Director, Science Monitoring Team, King's (formerly Chelsea) College, London
Professor D. Layton 1978–1982 Director, Science Monitoring Team, Leeds
Mr D. T. E. Marjoram HMI 1978–1979 HM Inspectorate
Mr I. W. Milligan HMI 1981–1986 Department of Education for Northern Ireland
Mr C. Parson HMI 1978–1983 HM Inspectorate
Dr B. M. Prestt 1978–1985 Manchester Polytechnic
Mr J. C. Taylor 1978–1982 Manchester LEA
Mr K. Wild 1978–1982 Staffordshire LEA
Mr G. D. Williams 1978–1982 Wellesbourne County Primary, Liverpool

A9.3 Monitoring Services Unit (NFER)

Mrs B. Bloomfield	Head of Unit
Mrs A. Baker	Deputy
Mrs E. Elliot	
Mrs M. Hall	
Mrs B. Woodley	
Mrs J. Cowan	Secretary

A9.4 APU Consultative Committee

Professor J. Dancy (Chair)	School of Education, University of Exeter
Miss J. E. L. Baird	Joint General Secretary AMMA
Dr P Biggs	Senior Adviser, Wiltshire LEA
Mrs M. J. Bloom	Project Leader for Building and Civil Engineering, National Economic Development Office
Mr P. Boulter	Director of Education, Cumbria (ACC)
Dr C. Burstall	Director, National Foundation for Educational Research
Professor C. B. Cox	Department of English Language and Literature, University of Manchester
Mrs J. Davies	Howbury Grange School, Bexley
Mr G. Donaldson	Flint High Comprehensive School (NUT)
Mr I. Donaldson	NAS/UWT
Mr H. Dowson	Deputy Headmaster, Earl Marshall School, Sheffield (NUT)
Mr G. Driver	Councillor, Leeds City Council (AMA)
Professor S. J. Eggleston	Department of Education, University of Keele
Mr A. Evans	Education Department, NUT
Mr D. Fox	Accountant, Chairman of National Education Association
Mr C. Gittins	Longsands School, St Neots (SHA)
Dr A. Grady	Middlesex Polytechnic
Mr P. L. Griffin	Windsor Clive Junior School, (NUT)
Mr K. S. Hopkins	Director of Education, Mid-Glamorgan (WJEC)
Mr C. Humphrey	Director of Education, Solihull (AMA)
Mr S. A. Josephs	MacMillan Education Limited
Mr J. A. Lawton	Kent County Council (ACC)
Mr G M Lee	Doncaster Metropolitan Institute of Higher Education (NATFHE)
Mr J. M. Leonard	General Inspector, Walsall LEA (AMA)
Mr M. J. Pipes	Headmaster, City of Portsmouth School for Boys (NAHT)
Mr G. R. Potter	Director of Education, West Sussex LEA (ACC)
Miss C. L. Richards	(CBI) Understanding British Industry Project, Birmingham
Mr R. Richardson	Advisory Head, ILEA (NUT)
Professor M. D. Shipman	School of Education, Roehampton Institute
Mr P. Smith	Springfield Lower School, Bedford
Mr S. C. Woodley	The Kings School, Canterbury

Assessors

Mr A. Gibson	HM Inspectorate
Mr K. A. Smart	Department of Education for Northern Ireland
Mr N. Summers	DES
Mr D. Timlin	Welsh Office Education Department

A9.5 APU Advisory Group on Statistics (September 1987)

Mr M. D. Phipps (Chair)	Administrative Head of the APU
Professor V. Barnett	Department of Statistics, University of Sheffield
Professor D. J. Bartholomew	Department of Statistics, London School of Economics and Political Science
Mrs B. Bloomfield	National Foundation for Educational Research
Mr T. Christie	Department of Education, University of Manchester
Mr J. Gardner	Chief Statistician, DES
Mr D. Hutchinson	Chief Statistician, NFER
Mrs S. Johnson	Centre for Studies in Science Education, University of Leeds
Professor T. Lewis	Faculty of Mathematics, Open University
Professor R. Mead	Department of Applied Statistics, University of Reading
Mr A. Owen	HMI

Continued overleaf

A9.5 APU Advisory Group on Statistics (September 1987)—continued

Mrs V. Scott	WOED
Dr A. S. Willmott	University of Oxford Delegacy of Local Examinations

A9.6 APU Management Group (September 1987)

Mr M. D. Phipps	Administrative Head of the APU
Mr A. G. Clegg HMI	Professional Head of the APU
Mr P. J. Silvester HMI	
Mr M. E. Malt	
Mr D. Sleep	
Miss H. Bennett	
Mrs M. L. Pooley	
Miss N. E. Mitchell	
Miss T. E. Pilborough	

Appendix

10

APU reports and other publications by science team members

Survey reports

1980 Surveys:

HARLEN W., BLACK P. and JOHNSON S. *Science in schools. Age 11: Report No 1*. London: HMSO, 1981

SCHOFIELD B., MURPHY P., JOHNSON S. and BLACK P. *Science in schools. Age 13: Report No 1*. London: HMSO, 1982

DRIVER R., GOTT R., JOHNSON S., WORSLEY C. and WYLIE F. *Science in schools. Age 15: Report No 1*. London: HMSO, 1982

1981 Surveys:

HARLEN W., BLACK P., JOHNSON S. and PALACIO D. *Science in schools. Age 11: Report No 2*. London: APU, 1983

SCHOFIELD B., BLACK P., HEAD J. and MURPHY P. *Science in schools. Age 13: Report No 2*. London: APU, 1984

DRIVER R., CHILD D., GOTT R., HEAD J., JOHNSON S., WORSLEY C. and WYLIE F. *Science in schools. Age 15: Report No 2*. London: APU, 1984

1982 Surveys:

HARLEN W., BLACK P., JOHNSON S., PALACIO D. and RUSSELL T. *Science in schools. Age 11: Report No 3*. London: APU, 1984

GOTT R., SCHOFIELD B., DAVEY A., GAMBLE R., HEAD J., KHALIGH N., MURPHY P., ORGEE T. and WELFORD G. *Science in schools. Ages 13 and 15: Report No 3*. London: APU, 1985

1983 Surveys:

HARLEN W., BLACK P., KHALIGH N., PALACIO D. and RUSSELL T. *Science in schools. Age 11: Report No 4*. London: APU, 1985

SCHOFIELD B., BLACK P., KHALIGH N., MURPHY P. and ORGEE T. *Science in schools. Age 13: Report No 4*. London: APU, 1986

WELFORD G., BELL J., DAVEY A., GAMBLE R. and GOTT R. *Science in schools. Age 15 Report No 4*. London: APU, 1986

1984 Surveys:

Included in the Review reports

Review reports

RUSSELL T., BLACK P., HARLEN W., JOHNSON S. and PALACIO D. *Science at age 11: A review of APU findings 1980–1984*. London: HMSO, 1988

SCHOFIELD B., BLACK P., BELL J. F., JOHNSON S., MURPHY P., QUALTER A. and RUSSELL T. *Science at age 13: A review of APU findings 1980–1984*. London: HMSO, 1988

ARCHENHOLD W. F., BELL J. F., DONNELLY J., JOHNSON S. and WELFORD G. *Science at age 15: A review of APU findings 1980–1984*. London: HMSO, 1988

Technical report

JOHNSON S. *National assessment: the APU science approach*. London: HMSO, 1988.

Science reports for teachers

HARLEN W. *Science at age 11*. Science Report for Teachers: 1. London: DES, 1983

MURPHY P. and GOTT R. *The assessment framework for science at ages 13 and 15*. Report for Teachers: 2. London: DES, 1984

MURPHY P. and SCHOFIELD B. *Science at age 13*. Science Report for Teachers: 3. London: DES, 1984

HARLEN W., PALACIO D. and RUSSELL T. *The APU assessment framework for science at age 11*. Science Report for Teachers: 4. London: DES, 1984

GOTT R. *Electricity at age 15*. Science Report for Teachers: 7. London: DES, 1984

GAMBLE R., DAVEY A., GOTT R. and WELFORD G. *Science at age 15*. Science Report for Teachers: 5. London: DES, 1985

WELFORD G., HARLEN W. and SCHOFIELD B. *Practical testing at ages 11, 13 and 15*. Science Report for Teachers: 6. London: DES, 1985

HARLEN W. *Planning scientific investigations at age 11*. Science Report for Teachers: 8. London: DES, 1987

GOTT R. and MURPHY P. *Assessing investigations in science, ages 13 and 15*. Science Report for Teachers: 9. London: DES, 1987

DONNELLY J. *Metals at Age 15*. Science Report for Teachers: 10. London: DES, 1988

WHITE J. and WELFORD G. *The Language of Science*. Report for Science Teachers: 11. London: DES, 1988

Articles, chapters and occasional papers

DRIVER R. and LAYTON D. 'A scientific approach to testing', *Education, 152*, 610–611, 1978

DRIVER R. and WORSLEY C. 'The Assessment of Performance in Science project', *European Journal of Science Education, 1*, 441–447, 1979

HARLEN W. 'Assessment of Performance in Science' in Fairbrother R, W, (Ed), *Assessment and the Curriculum*, Chelsea College, University of London, 1980

JOHNSON S. and MAHER B. 'Monitoring science performance using a computerized question-banking system', *British Journal of Educational Technology, 13*, 97–106, 1982

HARLEN W. 'Process skills, concepts and national assessment in science', *Research in Science Education, 13*, 1983

HARLEN W. 'The APU Survey–what scientific skills do 11 year olds have?' in Richards and Holford (Eds), *The Teaching of Primary Science: Policy and Practice.* London: Falmer Press, 1983

JOHNSON S., MURPHY P., DRIVER R., HEAD J. and PALACIO D. 'The science performances of boys and girls aged 11 to 15' in *Contributions to the second GASAT Conference.* Oslo: University of Oslo Institute of Physics, 1983

BLACK P., HARLEN W. and ORGEE T. 'Standards of performance–expectations and reality', *Journal of Curriculum Studies, 16*, 94–96, 1984

BLACK P., HARLEN W. and ORGEE T. 'Standards of performance–expectations and reality', *APU Occasional Paper No 3.* London: Assessment of Performance Unit, 1984

DRIVER R., HEAD J. and JOHNSON S. 'The differential uptake of science in schools in England, Wales and Northern Ireland', *European Journal of Science Education, 6*, 19–29, 1984

GAMBLE R. 'Data interpretation and representation in science' in *Contributions to University of Leeds/SSCR Conference 'Learning, Doing and Understanding'.* SSCR London, 1984

GOTT R. 'The APU results and their implications for curriculum development' in *Contributions to University of Leeds/SSCR Conference 'Learning, Doing and Understanding'.* SSCR London, 1984

HARLEN W. 'The impact of APU science work at LEA and school levels', *Journal of Curriculum Studies, 16*, 89–94, 1984

HARLEN W. 'Some results from the APU science surveys at age 11', *Primary Education Review, 19*, 12–14, 1984

JOHNSON S. and MAHER B. 'A thesaurus-linked science question-banking system', *British Journal of Educational Technology, 15*, 14–23, 1984

JOHNSON S. and MURPHY P. 'The underachievement of girls in physics: towards explanations', *European Journal of Science Education, 6*, 399–409, 1984

MURPHY P. 'Problem-Solving' in *Contributions to the University of Leeds/SSCR Conference 'Learning, Doing and Understanding in Science'.* SSCR London, 1984

WELFORD G. 'Assessment of practical skills' in *Contributions to the University of Leeds/SSCR Conference 'Learning, Doing and Understanding'.* SSCR London, 1984

BELL J. F., 'Generalizability theory: the software problem', *Journal of Educational Statistics, 10*, 19–29, 1985

GOTT R. 'The place of electricity in the assessment of performance in science–basic skills' in Duit, Jung and Rhoneck (Eds), *Aspects of Understanding Electricity–Proceedings of International Workshop*, Verlag, Schmidt & Klaunig, Kiel, 1985

GOTT R. 'Predicting and explaining the operation of simple DC cirucits' in Duit, Jung and Rhoneck (Eds), *Aspects of Understanding Electricity–Proceedings of International Workshop*, Verlag, Schmidt & Klaunig, Kiel, 1985

HARLEN W. 'Science at age 11' in Hodgson B. & Scanlan E (Eds), *Approaching Primary Science.* London: Harper and Row, 1985

JOHNSON S. and BELL J. F. 'Evaluating and predicting survey efficiency using generalizability theory', *Journal of Educational Measurement, 22*, 107–119, 1985

DONNELLY J. F. and GOTT R. 'An assessment-led approach to processes in the science curriculum', *European Journal of Science Education, 7*, 237–251, 1985

JOHNSON S. and MURPHY P. 'Girls and physics: reflections on APU survey findings', *APU Occasional Paper No 4.* London: Asssessment of Performance Unit, 1986

GAMBLE R. 'Simple equations in physics', *European Journal of Science Education, 8*, 27–37, 1986

MURPHY P. 'Differences between girls and boys in the APU science results', *Primary Science Review No 2,* Autumn 1986

BELL J. F. 'Simultaneous confidence intervals for the linear functions of expected mean squares used in generalizability theory', *Journal of Educational Statistics, 11*, 3, 197–205, 1986

MURPHY P. 'The APU Science INSET kits – some possible uses', *Education in Science, No 1,* January 1987

DONNELLY J. F. '15 year old pupils variable handling performance in the context of scientific investigations', *Research in Science & Technological Education, 5*, 2, 135–147, 1987

JOHNSON S. 'Gender differences in science: parallels in interest, experience and performance', *International Journal of Science Education, 9*, 4, 467–481, 1987

JOHNSON S. & BELL J. F. 'Gender differences in science: option choices', *School Science Review, 69*, 247, 268–276, 1987

JOHNSON S. 'Assessment in science and technology', *Studies in Science Education, 14,* 83–108, 1987

JOHNSON S. 'Early developed sex differences in science and mathematics', *Journal of Early Adolescence, 7,* 1, 21–33, 1987

JOHNSON S. and BELL J. 'Gender differences in science: option choices', *School Science Review, 69,* 247, 266–276, 1987

GOTT R. and WELFORD G. 'The assessment of observation in science', *School Science Review, 69,* 247, 217–227, 1987

DONNELLY J. F. and WELFORD A. G. 'Children's performance in chemistry', *Education in Chemistry, 25,* 1, 7–10, 1988

Index

Printed in the United Kingdom for Her Majesty's Stationery Office
Dd239893 5/88 C38 G443 10170